United States Nuclear Regulatory Commission

Protecting People and the Environment

NUREG-1873

I0488209

Environmental Impact Statement for License Renewal of the National Bureau of Standards Reactor

Final Report

Manuscript Completed: October 2007
Date Published: December 2007

Office of Nuclear Reactor Regulation

Abstract

This environmental impact statement (EIS) was prepared in response to an application submitted to the U.S. Nuclear Regulatory Commission (NRC) by the National Institute of Standards and Technology (NIST) to renew the operating license for the National Bureau of Standards Reactor (NBSR) for a period of an additional 20 years. This is the second operating license renewal application for the NBSR. The first license renewal was granted May 16, 1984, and included a power uprate from 10 megawatts of thermal power (MWt) to 20 MWt. This EIS includes the NRC staff's analysis that considers and weighs the environmental impacts of the proposed action, as well as mitigation measures available for reducing or avoiding adverse impacts. It also includes the staff's recommendation regarding the proposed action.

No public comments were received during the scoping process. Two comment letters – one from the U.S. Department of the Interior and one from the U.S. Environmental Protection Agency – were received during the comment period provided for review of the Draft EIS. These comment letters are provided in Appendix B, part II of this Final EIS. The staff determined from its review of the application that no issues having a significant environmental impact exist, and additional mitigation measures are not likely to be sufficiently beneficial as to be warranted.

The NRC staff's recommendation is that the Commission determines the adverse environmental impacts of license renewal for the NBSR are not so great that license renewal would be unreasonable. This recommendation is based on (1) the Environmental Report submitted by NIST; (2) consultation with Federal, State, and local agencies; and (3) the staff's own independent review.

Paperwork Reduction Act Statement

Public Protection Notification

Contents

Contents

Contents

Contents

Figures

Tables

Executive Summary

By letter dated April 9, 2004, the National Institute of Standards and Technology (NIST) submitted an application to the U.S. Nuclear Regulatory Commission (NRC) to renew the operating license (OL) for the National Bureau of Standards Reactor (NBSR) for an additional 20-year period. This is the second license renewal application for the NBSR. The first license renewal was granted May 16, 1984, and included a power uprate from 10 megawatts of thermal power (MWt) to 20 MWt. If the current OL is not renewed, then the reactor must be shut down.

Section 102 of the National Environmental Policy Act (NEPA) (42 USC 4321 et seq.) directs that an environmental impact statement (EIS) is required for major Federal actions that could significantly affect the quality of the human environment. The NRC has implemented Section 102 of NEPA in Title 10 of the Code of Federal Regulations (CFR) Part 51. Part 51 identifies licensing and regulatory actions that require an EIS. In 10 CFR 51.20(b)(2), the Commission requires preparation of an EIS for renewal of a testing facility (test reactor) OL.

Upon acceptance of the NIST application, the NRC began the environmental review process described in 10 CFR Part 51 by publishing a Notice of Intent to prepare an EIS and conduct scoping (70 FR 56935) on September 29, 2005. The staff visited the NIST site in September 2006. In the preparation of this EIS for the NBSR, the staff reviewed the NIST Environmental Report (ER), consulted with other agencies, and conducted an independent analysis of the issues. No comments were received from the public during the scoping process.

A Draft EIS was published for comment in June 2007. On July 18, 2007, the NRC published a Notice of Availability for the Draft EIS, thus initiating a comment period that ended on September 5, 2007 (72 FR 39467). Two comments – one from the U.S. Department of the Interior and one from the U.S. Environmental Protection Agency – were received, and these comments are addressed in Appendix B, Part II, of this Final EIS.

This EIS includes the NRC staff's analysis that considers and weighs the environmental effects of the proposed action, the environmental impacts of alternatives to the proposed action, and mitigation measures for reducing or avoiding adverse effects. It also includes the staff's recommendation regarding the proposed action.

Executive Summary

For the evaluation of the NBSR license renewal action, the staff has applied the NRC's three-level standard of significance – SMALL, MODERATE, or LARGE – developed using the Council on Environmental Quality guidelines. The following definitions of the three significance levels are set forth in the footnotes to Table B-1 of 10 CFR Part 51, Subpart A, Appendix B:

SMALL – Environmental effects are not detectable or are so minor that they will neither destabilize nor noticeably alter any important attribute of the resource.

MODERATE – Environmental effects are sufficient to alter noticeably, but not to destabilize, important attributes of the resource.

LARGE – Environmental effects are clearly noticeable and are sufficient to destabilize important attributes of the resource.

The staff's analysis revealed that all of the environmental impacts considered in this EIS for continued operation of the NBSR during the term of the renewed OL would be expected to be SMALL.

If the NBSR operating license is not renewed and the unit ceases operation, then the adverse impacts of the most likely alternative, construction of a replacement facility, will not be smaller than those associated with continued operation of the NBSR. The impacts may, in fact, be greater in some areas.

The recommendation of the NRC staff is that the Commission determines the adverse environmental impacts of license renewal for the NBSR are not so great that preserving the option of license renewal for NIST decisionmakers would be unreasonable. This recommendation is based on (1) the ER submitted by NIST; (2) consultation with other Federal, State, and local agencies; and (3) the staff's own independent review. There were no comments received from the public during the scoping process or the comment period following publication of the Draft EIS that would require the NRC to consider additional environmental issues above those anticipated by the staff to be relevant.

Abbreviations/Acronyms

α	alpha
γ	gamma
μCi	microcurie(s)
ac	acre(s)
ADAMS	Agencywide Document Access and Management System
ALARA	as low as reasonably achievable
AQCR	Air Quality Control Region
AQI	Air Quality Index
BWI	Baltimore-Washington International Airport
°C	degrees Celsius
CAM	continuous air monitor
CEQ	Council on Environmental Quality
CFR	Code of Federal Regulations
cfs	cubic feet per second
Ci	curie(s)
cm	centimeter(s)
DAC	derived air concentration
DBA	design basis accident
DCA	Ronald Reagan Washington National Airport
DOE	U.S. Department of Energy
EIS	environmental impact statement
EPA	U.S. Environmental Protection Agency
EPZ	emergency planning zone
ER	Environmental Report
ESF	Engineered Safety Features
°F	degrees Fahrenheit
FES	Final Environmental Statement
FR	*Federal Register*
ft	foot/feet

gal	gallon
GEIS	Generic Environmental Impact Statement for License Renewal of Nuclear Plants, NUREG-1437
gpd	gallon(s) per day
gpm	gallon(s) per minute
ha	hectare(s)
HEPA	high-efficiency particulate air (filter)
HEU	highly enriched fuel
HLW	high-level waste
hr	hour(s)
IAD	Dulles International Airport
in.	inch(es)
kg	kilogram(s)
km	kilometer(s)
L	liter(s)
lb	pound(s)
LEU	low enriched fuel
LLW	low-level radioactive waste
LWR	light-water reactor
m	meter(s)
MGD	million gallons per day
MHA	maximum hypothetical accident
mi	mile(s)
min	minute(s)
mL	milliliter(s)
MLLW	mixed low level waste
mm	millimeters
mrem	millirem(s)
m/s	meters per second
mSv	millisievert(s)
MTR	materials testing reactor
MW	megawatt(s)
MWe	megawatt(s) electric
MWt	megawatt(s) thermal

Abbreviations/Acronyms

n	neutron
NAAQS	National Ambient Air Quality Standards
NBSR	National Bureau of Standards Reactor
NEPA	National Environmental Policy Act of 1969, 42 USC 4321, et seq.
NHPA	National Historic Preservation Act of 1966, 16 USC 470, et seq.
NIST	National Institute of Standards and Technology
NRC	U.S. Nuclear Regulatory Commission
NTSB	National Transportation Safety Board
NWS	National Weather Service
OL	operating license
OSTP	Office of Science and Technology Policy
p	proton
rem	special unit of dose equivalent, equal to 0.01 Sv
REMP	radiological environmental monitoring program
s	second(s)
SAR	safety analysis report
SER	Safety Evaluation Report
Sv	sievert (special unit of dose equivalent)
TLD	thermoluminescent dosimeter
USC	United States Code
USCB	U.S. Census Bureau
WSSC	Washington Suburban Sanitary Commission
yr	year(s)

1.0 Introduction

Under the U.S. Nuclear Regulatory Commission's (NRC's) environmental protection regulations in Title 10 of the Code of Federal Regulations (CFR) Part 51, which implement the National Environmental Policy Act of 1969 (NEPA), renewal of a nuclear test reactor operating license (OL) requires the preparation of an environmental impact statement (EIS). In preparing the EIS, the NRC staff is required first to issue the statement in draft form for comment, and then issue a final statement after considering comments on the draft.

The National Institute of Standards and Technology (NIST) operates the National Bureau of Standards Reactor (NBSR) in Gaithersburg, Maryland, under OL No. TR-5, which was issued by the NRC. By letter dated April 9, 2004, NIST submitted an application to the NRC to renew the OL for the NBSR for an additional 20 year period under 10 CFR 51.20(b)(2). The reactor is a high-flux, heavy-water-moderated, cooled, and reflected test reactor that first went critical on December 7, 1967, after receiving a provisional OL. Facility Operating License TR-5, authorizing operation at a maximum power level of 10 megawatts of thermal power (MWt) for a period of 15 years, was issued on June 30, 1970 (U.S. NRC 1982). This is the second license renewal application for the NBSR. The first license renewal was granted May 16, 1984, and included a power uprate from 10 MWt to 20 MWt. Pursuant to 10 CFR 51.53, NIST submitted an Environmental Report (ER) (NIST 2004), which analyzed the environmental impacts associated with the proposed license renewal action and evaluated mitigation measures for reducing adverse environmental effects. The current OL for the NBSR was scheduled to expire on May 16, 2004. However, in accordance with 10 CFR 2.109(a) NIST's application for renewal was received at least 30 days prior to the expiration of an existing license, and therefore, the existing OL will not be considered expired until the application has been finally determined.

This report is the EIS for the NIST application for license renewal of the NBSR. The NRC staff will also prepare a separate safety evaluation report.

1.1 Report Contents

The following sections of this Introduction (1) describe the background for the preparation of this EIS and the process used by the staff to assess the environmental impacts associated with license renewal; (2) describe the proposed Federal action to renew the NBSR OL; (3) discuss the purpose and need for the proposed action; and (4) discuss the NBSR's compliance with environmental quality standards and requirements that have been imposed by Federal, State, regional, and local agencies that are responsible for environmental protection.

The ensuing chapters of this EIS include the following information. Chapter 2 describes the site, reactor, and interactions of the reactor with the environment. Chapter 3 discusses the environmental impacts of operation during the renewal term. Chapter 4 contains a summary of the evaluation of potential environmental impacts of plant accidents, including consideration of the maximum hypothetical event. Chapter 5 discusses the environmental impacts of the uranium fuel cycle and solid waste management. Chapter 6 examines the impacts of decommissioning. Chapter 7 discusses the impacts of alternatives to license renewal. Chapter 8 summarizes the findings of the preceding chapters and draws conclusions about the adverse impacts that cannot be avoided, the relationship between short-term uses of the environment and the maintenance and enhancement of long-term productivity, and any irreversible or irretrievable commitment of resources. Chapter 8 also presents the staff's preliminary recommendation with respect to the proposed license renewal action.

Additional information is included in appendixes. Appendix A lists the contributors to the document. Appendix B addresses comments received during the environmental review of the Draft EIS for license renewal. Appendix C provides a chronology of the NRC staff's environmental review correspondence related to this EIS, and Appendix D identifies the organizations contacted during the development of the EIS.

1.2 Background

An applicant seeking to renew its OL is required to submit an ER as part of its application. The NRC license-renewal evaluation process involves careful (1) review of an applicant's ER; (2) review of records of comments received; (3) review of environmental quality standards and regulations; (4) coordination with Federal, State, and local environmental protection and resource agencies; and (5) review of the technical literature to verify the environmental impacts of the proposed license renewal. Using the NRC's established license renewal evaluation framework for commercial power reactors ensures a thorough evaluation of the impacts of renewal of the OL for the NBSR. The *Generic Environmental Impact Statement for License Renewal of Nuclear Plants (GEIS)*, NUREG-1437, Volumes 1 and 2 (U.S. NRC 1996, 1999),[a] was written specifically for use in the renewal of OLs for commercial power reactors. In conducting the staff review of the NIST application, the NRC staff was informed by certain GEIS features including the use of the three-level standard of significance, which is described below.

In following the precedent of the GEIS and the site-specific supplemental license renewal EISs, environmental issues in this EIS have been evaluated using a three-level standard of significance – SMALL, MODERATE, or LARGE – developed by NRC using guidelines from the Council on Environmental Quality. The definitions of the three significance levels are set forth in the footnotes to Table B-1 of 10 CFR Part 51, Subpart A, Appendix B, as follows:

(a)   The GEIS was originally issued in 1996. Addendum 1 to the GEIS was issued in 1999. Hereafter, all references to the "GEIS" include the GEIS and its Addendum 1.

SMALL – Environmental effects are not detectable or are so minor that they will neither destabilize nor noticeably alter any important attribute of the resource.

MODERATE – Environmental effects are sufficient to alter noticeably, but not to destabilize, important attributes of the resource.

LARGE – Environmental effects are clearly noticeable and are sufficient to destabilize important attributes of the resource.

When the findings in the GEIS are used in this document, there is a description of the finding and a brief discussion on how the findings can also be applicable to or bound the environmental effects of a test reactor such as the NBSR.

As a part of its review, the NRC prepares an independent analysis of the environmental impacts of license renewal and compares these impacts with the environmental impacts of alternatives. The evaluation of the NIST license renewal application began with the publication of a Notice of Acceptance for docketing and opportunity for a hearing in the *Federal Register* (69 FR 56462) on September 21, 2004. The staff issued a Notice of Intent to prepare a draft EIS and to conduct scoping (70 FR 56935) on September 29, 2005.

The NRC staff and contractors retained to assist the staff visited the NIST site on September 26, 2006, to gather information and to become familiar with the NBSR, the NIST site, and its environs. The staff also consulted with Federal, State, regional, and local agencies. A list of the organizations consulted is provided in Appendix D. Other documents related to the NBSR were reviewed and are referenced. There were no comments received from the public related to the NBSR during the scoping period, which ended November 28, 2005.

A Draft EIS was published for comment in June 2007. On July 18, 2007, the NRC published a Notice of Availability for the Draft EIS, thus initiating a comment period that ended on September 5, 2007 (72 FR 39467). Two comments – one from the U.S. Department of the Interior and one from the U.S. Environmental Protection Agency – were received, and these comments are addressed in Appendix B, Part II, of this Final EIS.

This EIS presents the staff's analysis that considers and weighs the environmental effects of the proposed renewal of the OL for the NBSR, the environmental impacts of alternatives to license renewal, and mitigation measures available for avoiding adverse environmental effects. Chapter 8, Summary and Conclusions, provides the staff's recommendation to the Commission on whether or not the adverse environmental impacts of license renewal are so great that preserving the option of license renewal would be unreasonable.

1.3 The Proposed Federal Action

The proposed Federal action is renewal of the OL for the NBSR. This reactor is located on the NIST campus in upper Montgomery County, Maryland, approximately 32 km (20 mi) northwest of the District of Columbia. The NBSR is a heavy water-moderated and cooled, enriched-fuel, tank-type reactor designed to operate at 20 megawatts of thermal power (MWt). It is a custom-designed variation of the Argonne CP-5 class reactor. The primary coolant system is closed, recirculating heavy water (D_2O) in an aluminum and stainless steel system. Heat from the reactor is transferred to a secondary cooling system of light water, and then to the atmosphere by means of evaporation from a cooling tower located outside the confinement building (NIST 2004). The current OL for the NBSR expired on May 16, 2004. By letter dated April 9, 2004, NIST submitted an application to the NRC to renew this OL for an additional 20 years of operation (NIST 2004). Because the license renewal application was filed in a timely manner under 10 CFR 2.109, the license is not deemed to have expired until a final determination has been made on the application.

1.4 The Purpose and Need for the Proposed Action

Although a licensee must have a renewed license to operate a reactor beyond the term of the existing OL, the possession of that license is just one of a number of conditions that must be met for the licensee to continue plant operation during the term of the renewed license.

The NIST Center for Neutron Research is a national resource used by up to 2000 engineers and scientists per year for research in materials science, non-destructive evaluation, chemistry, biology, trace analysis, neutron standards and dosimetry, nuclear physics, and quantum metrology. A large cold-neutron source (which slows neutrons to speeds of 1000 m/s or less and produces very low energy neutrons for research purposes) and seven neutron guides provide the United States with world-class capabilities in cold neutron research. The NBSR is used by engineers and scientists from all over the country, and is operated continuously (24 hours a day, 7 days a week), with routine shutdowns every 5 to 6 weeks for partial refueling and, as needed, for maintenance. A study by an interagency working group of the Office of Science and Technology Policy (OSTP 2002) stated that the NIST Center for Neutron Research was the highest performing neutron facility in the United States at that time.

Thus, for this license renewal review, the NRC considers the purpose and need for the proposed action (i.e., renewal of the NIST NBSR OL) is to provide an option allowing for neutron research capabilities beyond the term of the current reactor OL to meet national research and test facility needs, as such needs may be determined by NIST (and other Federal decisionmakers).

This definition of purpose and need reflects the Commission's recognition that, unless there are findings in the safety review required by the Atomic Energy Act of 1954 or in the NEPA environmental analysis that would lead the NRC to reject this license-renewal application, the NRC does not have a role in the research-planning decisions as to whether this reactor should continue to operate. From the perspective of the licensee, the purpose of renewing this OL is to maintain the availability of specific research capabilities beyond the current term of NIST's OL.

1.5 Compliance and Consultations

The NBSR uses municipal water for cooling and discharges used cooling water into the municipal sewer in accordance with the NIST discharge permit. The NIST campus is not within Maryland's coastal zone; therefore, the site is not subject to the Coastal Zone Management Act. NRC staff consulted with the Maryland Historical Trust regarding the potential renewal of the OL for the NBSR and determined, in accordance with 36 CFR 800.3(a)(1), that renewal would be an activity that does not have the potential to cause effects on historic properties.

Emergency power generators and other facilities and activities associated with the NIST site emit various pollutants, which are regulated under Title V Operating Permit 24-030-00323 by the Maryland Department of the Environment, Air Quality Permits Program, Air and Radiation Management Administration; the permit is scheduled to expire on April 30, 2008.

Section 7(a)(2) of the Endangered Species Act states that Federal agencies are to consult with the U.S. Fish and Wildlife Service (FWS) to ensure that any agency action is not likely to jeopardize the continued existence of any endangered species or threatened species or result in the destruction or adverse modification of habitat of such species. Although no threatened or endangered species are known to occur on the NIST campus, official consultation was initiated with the FWS.

1.6 References

10 CFR Part 51. Code of Federal Regulations, Title 10, *Energy,* Part 51, "Environmental Protection Regulations for Domestic Licensing and Related Regulatory Functions."

69 FR 56462. "Notice of Acceptance for Docketing of the Application and Notice of Opportunity for a Hearing Regarding Renewal of the National Bureau of Standards Reactor (The NBSR) Facility Operating License No. TR5 for an Additional Twenty-Year Period." *Federal Register,* Vol. 69, No. 182, pp. 56,462-56,464. September 21, 2004.

Introduction

70 FR 56935. "National Institute of Standards and Technology, National Bureau of Standards Reactor; Notice of Intent to Prepare an Environmental Impact Statement and Conduct Scoping Process." *Federal Register*, Vol. 70, No. 188, pp. 56,935-56,936. September 29, 2005.

72 FR 39467. 2007. "National Institute of Standards and Technology; National Bureau of Standards Reactor; Notice of Availability of the Draft Environmental Impact Statement for License Renewal and Public Comment Period for the License Renewal of National Bureau of Standards Reactor." *Federal Register*, Vol. 72, No. 137, p. 39467. July 18, 2007.

Atomic Energy Act of 1954 (AEA). 42 USC 2011, et seq.

Coastal Zone Management Act (CZMA). 16 USC 1451, et seq.

Endangered Species Act of 1973. 16 USC 1531, et seq.

National Environmental Policy Act of 1969 (NEPA). 42 USC 4321, et seq.

National Historic Preservation Act of 1966. 16 USC 470, et seq.

National Institute of Standards and Technology (NIST). 2004. *Environmental Report for License Renewal for the National Institute of Standards and Technology Reactor – NBSR*. Docket No. 50-184, License No. TR5, Gaithersburg, Maryland. ML041120176.

National Institute of Standards and Technology (NIST). 2004. Letter dated April 9, 2004 from Seymore H. Weiss, Chief Reactor Operations and Engineering. Subject: License Renewal Application for the National Institute of Standards and Technology Reactor. ML041120167.

Office of Science and Technology Policy (OSTP). 2002. *Report on the Status and Needs of Major Neutron Scattering Facilities and Instruments in the United States*. Washington, D.C.

U.S. Nuclear Regulatory Commission (U.S. NRC). 1982. *Final Environmental Impact Statement Related to License Renewal and Power Increase for the National Bureau of Standards Reactor*. NUREG-0877, Washington, D.C.

U.S. Nuclear Regulatory Commission (U.S. NRC). 1996. *Generic Environmental Impact Statement for License Renewal of Nuclear Plants*. NUREG-1437, Volumes 1 and 2, Washington, D.C.

U.S. Nuclear Regulatory Commission (U.S. NRC). 1999. *Generic Environmental Impact Statement for License Renewal of Nuclear Plants, Main Report*, "Section 6.3 – Transportation, Table 9.1, Summary of findings on NEPA issues for license renewal of nuclear power plants, Final Report." NUREG-1437, Volume 1, Addendum 1, Washington, D.C.

2.0 Description of Reactor, Site, and Reactor Interaction with the Environment

The National Institute of Standards and Technology (NIST) Center for Neutron Research is a reactor-laboratory complex providing NIST and the nation with a facility for the performance of neutron-based research. The heart of this facility is the National Bureau of Standards Reactor (NBSR). The facility is located on the 234.5-ha (579.5-ac) NIST campus in upper Montgomery County, Maryland, approximately 32 km (20 mi) northwest of the District of Columbia (U.S. NRC 2007). NIST is a non-regulatory Federal agency of the U.S. Commerce Department within the Technology Administration.

The NIST Center for Neutron Research is a national resource used by nearly 2000 engineers and scientists each year. In 2002, researchers came to the center from all areas of the country, including 30 other Federal laboratories, 127 universities, 47 industrial laboratories, and 21 NIST divisions and offices. The major research areas include materials science, non-destructive evaluation, chemistry, biology, trace-element analysis, neutron standards and dosimetry, nuclear physics, and quantum metrology. A large cold-neutron source and seven neutron guides provide the United States with capabilities in cold-neutron research, and up to 25 cold and thermal neutron instruments provide neutron scattering capability. As a result, the Center for Neutron Research served over 60 percent of the neutron users in the United States during the period 2000 through 2003. The reactor is operated 24 hours per day, 7 days per week, which allows for the operation of an extensive user program (NIST 2004).

Unless otherwise indicated, information in the following sections was adapted from the Environmental Report (ER) submitted by NIST for renewal of the NBSR operating license (OL) (NIST 2004a), and was independently verified by the staff. Additional information was obtained by the staff during the site audit (U.S. NRC 2007); appropriate citations will be made for other sources. The plant and its environment are described in Section 2.1, interactions of the plant with the environment are presented in Section 2.2, and references are listed in Section 2.3.

2.1 Reactor and Site Description and Proposed Reactor Operation During the License Renewal Term

The NIST Center for Neutron Research reactor-laboratory complex provides NIST and the nation with an extensive facility for neutron-based research in biology, chemistry, engineering, materials science, and physics.

2.1.1 External Appearance and Setting

NIST is located within the Interstate-270 (I-270) Technology Corridor, as shown in Figures 2-1 and 2-2. This corridor is sited strategically in the center of Montgomery County and constitutes the county's primary focus of economic and transportation activity. The corridor straddles I-270 from the I-495 Washington Beltway to the south, to Clarksburg on the north. Figure 2-3 provides an overview of the NIST campus, and Figure 2-4 shows the layout of the NIST Center for Neutron Research reactor-laboratory complex in Building 235.

The site is suitable for the NBSR, given the reactor's characteristics (see Section 2.1.2). In particular, it operates at low power, at near-atmospheric pressure, and at low temperature. Consequently, there is neither a large inventory of radioactive fission products nor stored thermal energy to disperse that inventory to the surrounding area. The NBSR facility also has a confinement building to limit any radiological release to the environment in the unlikely event of an accident.

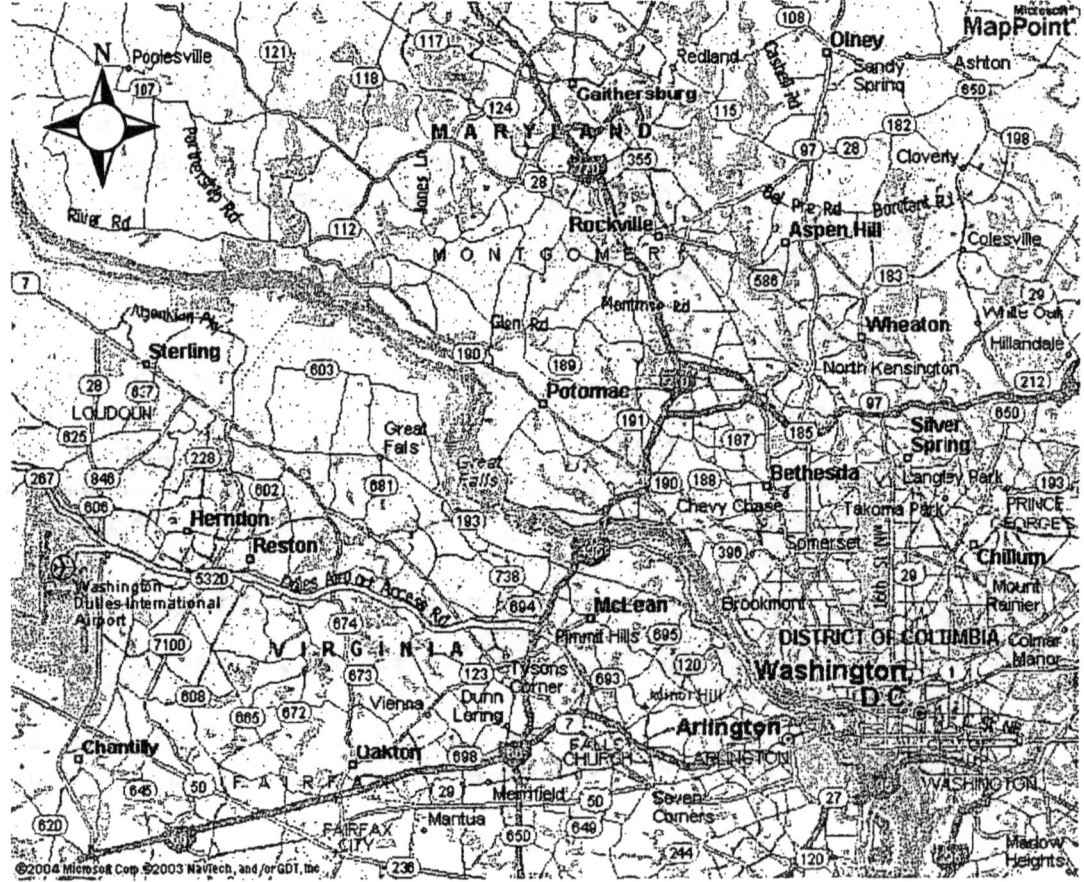

Figure 2-1. Regional Map

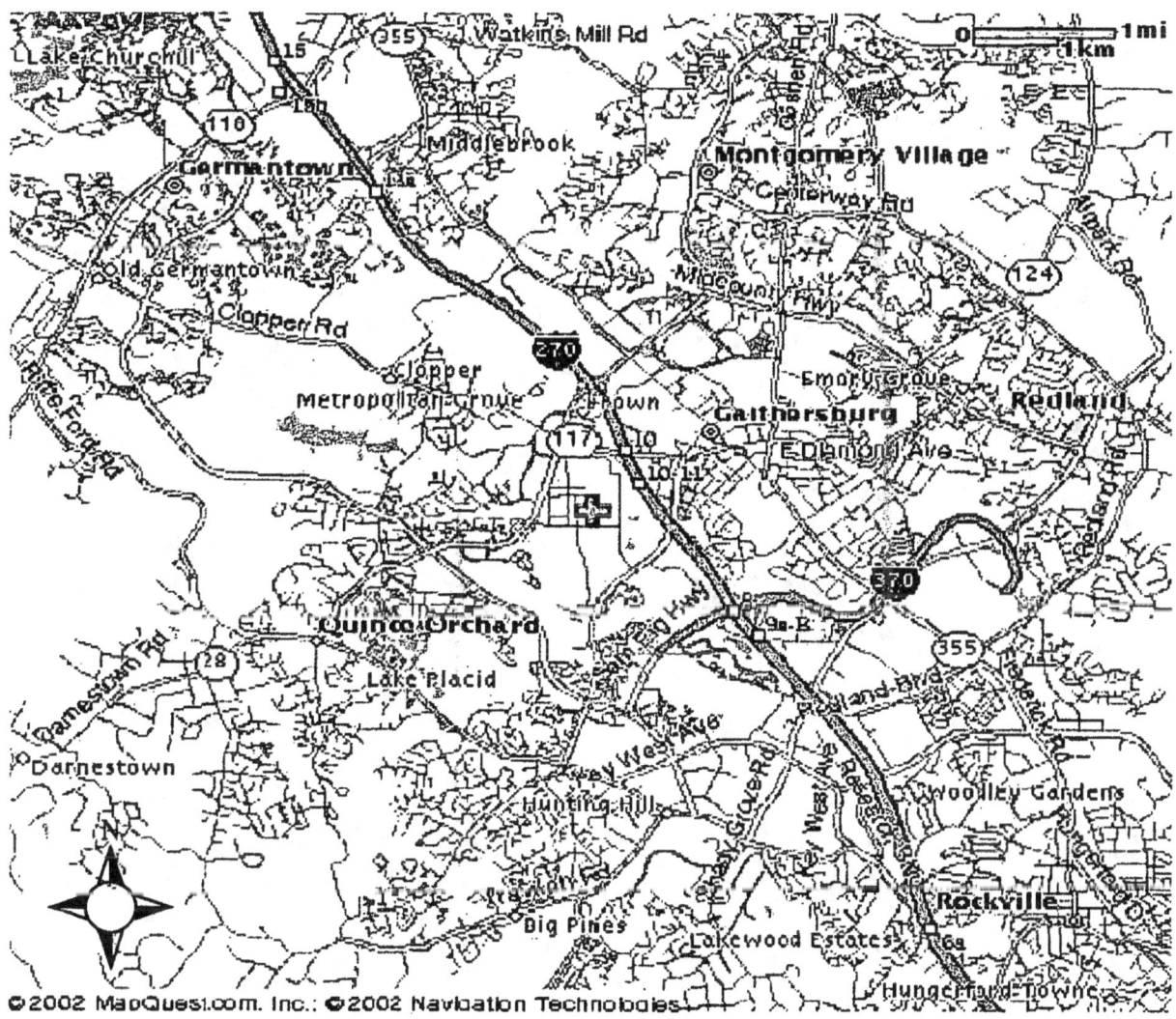

Figure 2-2. NIST Immediate Area

Figure 2-3. NIST Photographic View

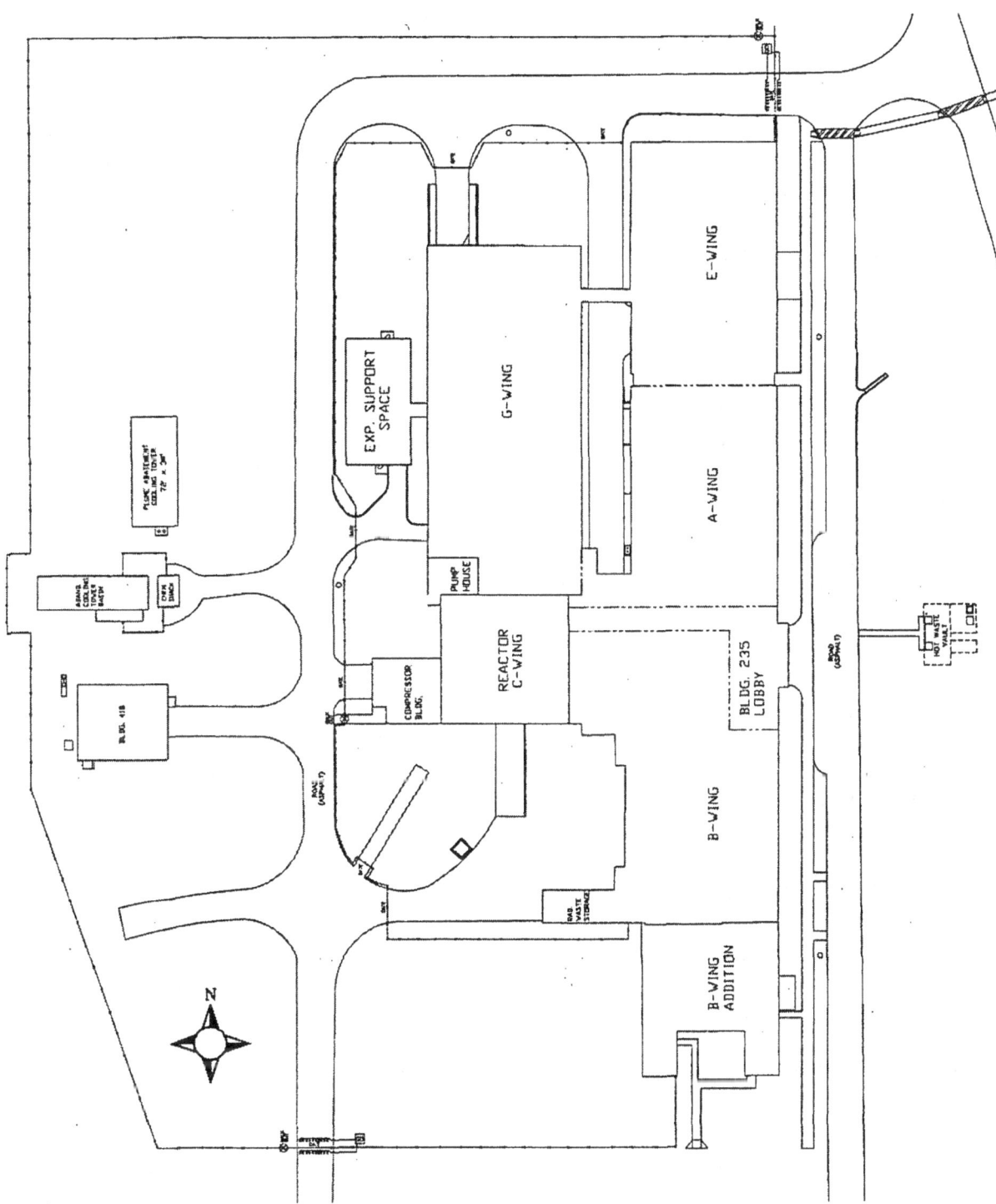

Figure 2-4. NIST Center for Neutron Research

Reactor and the Environment

2.1.1.1 Specification and Location

The NIST Center for Neutron Research is located on the 234.5-hectare (579.5-acre) NIST campus (U.S. NRC 2007) in upper Montgomery County, approximately 32 km (20 mi) northwest of Washington, D.C. (Figure 2-1). The reactor-laboratory complex is located on Center Drive in the southern portion of the NIST campus in Gaithersburg, Montgomery County, Maryland (Figures 2-2, 2-3, and 2-4). There are no prominent natural features in the immediate vicinity of the reactor, and the most prominent man-made feature is I-270 adjacent to the eastern boundary of the NIST campus.

2.1.1.2 Access Control and Emergency Planning Zone

Only portions of the Center for Neutron Research facility in Building 235 are directly affected by the U.S. Nuclear Regulatory Commission (NRC) license: those include parts of the confinement building in C-Wing under licensed operations, the Guide Hall and its auxiliary building in the Cold Neutron Guide Hall in G-Wing, the ventilation stack east of the pump house, the emergency control station (ECS) and the fuel storage area (FSA) located in the A-Wing basement area, the heating, ventilation and air conditioning (HVAC) and electrical service equipment in the B-Wing basement, and the high-bay area located on the main level of the B-Wing immediately adjacent to the east side of the confinement building.

There are a number of access controls related to the reactor:

- The NIST boundary fence, which surrounds the campus – Access is controlled by NIST Security, and access is limited to employees, contractors, and individuals who have business onsite. This includes the NIST Child Care Center, which lies within 1 km (0.6 mi) of the reactor.

- The NBSR site boundary, which is marked by the perimeter fence that surrounds Building 235, including the nearby cooling towers, the chemical building, and Building 418, which includes a radioactive waste storage and shipment building in the H-Wing. Within this area, unescorted access is limited to those individuals on the access list; all others require an escort.

- The reactor operations boundary, which coincides with the building perimeter – This includes the G-Wing and its auxiliary support building (compressor building for cold neutron cryostat in F-Wing and the experiment support space in J-Wing), the office areas and support spaces in E-Wing, and the radioactive waste storage area west of B-Wing.

The Emergency Planning Zone (EPZ) is marked by a 400-m (0.25-mi) radius centered on the ventilation stack. There is no public access to the EPZ, which is located entirely within the NIST campus, and access is limited to individuals having business there. The NIST Child Care Center lies outside the EPZ.

2.1.1.3 Population Distribution

Because the NBSR lies entirely on the NIST campus, the area immediately around the reactor contains laboratories and office buildings but no residential buildings and no part-time, transient, or seasonal residents. Permanent residences are at least 400 m (0.25 mi) directly to the east and the west of the reactor.

Populations within the 1-, 2-, 4-, 6-, and 8-km (0.6-, 1.2-, 2.5-, 3.7-, and 5-mi) radii around the reactor were estimated from the 2000 Census population counts by jurisdiction for the voting districts located within these areas. Table 2-1 provides current populations within the five radii identified above for the year 2000, based on the voting district data, as well as projections for the population in 2010 and in 2025. These values were derived by applying the percentage changes, as determined from the Montgomery County planning area forecasts listed in 2000 Census data (USCB 2000). For voting districts that cross into more than one of the zones around the NBSR, the percentage area located within each ring was estimated, and the population distribution within any one district was assumed to be in proportion to the area.

NIST and the NBSR lie within Montgomery County, which is the most populous county in the State of Maryland. Table 2-2 provides the 1950 to 2000 Census Population and percentage changes for the County. Table 2-3 lists population forecasts from 2000 through 2025 for Montgomery County, as provided by the National Capital Park and Planning Commission – Montgomery County Planning Board, and Table 2-4 provides the Montgomery County Planning Area forecasts for 2005 to 2025.

Table 2-1. Population Estimates Around the NIST Campus

Circle Radii (km)	2000	2010	2025
1	3462	3677	4054
2	19,178	20,367	22,457
4	73,121	77,654	85,624
6	155,402	168,163	180,247
8	218,752	237,848	253,100

Table 2-2. Montgomery County Population

Year	Population	Percentage Change
1950	164,401	n/a
1960	340,928	107.4
1970	522,809	62.0
1980	579,053	10.8
1990	757,027	30.7
2000	873,341	15.4

Table 2-3. Montgomery County Population Forecasts

Year	Population	Percentage Change
2000	873,341	n/a
2005	925,000	6.0
2010	975,000	5.4
2015	1,020,000	4.6
2020	1,050,000	2.9
2025	1,070,000	1.9

Table 2-4. Montgomery County Planning Area Forecasts for Population

Planning Area	Year				
	2005	2010	2015	2020	2025
Darnestown	12,9000	13,300	13,900	14,600	14,600
Gaithersburg	125,400	127,900	133,300	139,000	141,000
Germantown	81,000	82,300	85,600	86,800	86,800
Potomac	44,800	46,000	47,800	49,600	50,200
Rockville	48,900	52,500	51,000	50,100	50,000

Surrounding the NIST campus is the city of Gaithersburg, which encompasses all of the 2-km (1.25-mi) radius around the reactor and most of 4-km (2.5-mi) radius. All of the town of Washington Grove and much of the city of Rockville lie within the 8-km (5-mi) circle. Other unincorporated areas of Montgomery County within an 8-km (5-mi) radius are Germantown, Montgomery Village, Darnestown, and North Potomac. According to the 2000 Census, the Germantown area was the seventh most populous community in Maryland with 55,419 residents; Gaithersburg was tenth with 52,613; Rockville was fourteenth at 47,388; and Montgomery Village was twenty-first at 38,051. In terms of percentage growth of their populations between 1990 and 2000, this represents an increase of 35, 33, 5.7, and 18 percent, respectively. Table 2-5 presents the 1990 and 2000 Census Data for these communities.

Table 2-5. NBSR Site Area Census Data

	1990 Population	2000 Population
Gaithersburg	39,542	52,613
Rockville	44,835	47,388
Washington Grove	--	515
Germantown	41,145	55,419
Montgomery Village	32,315	38,051
North Potomac	–	23,044
Darnestown	–	6378

2.1.1.4 Nearby Industrial, Transportation, and Military Facilities

NIST is located between several major roads, with Interstate Highway 270 (I-270) at the northeast boundary. I-270 is a major commuter and truck route between northern Montgomery County, Frederick County, and other points north to the employment areas in the Washington, D.C., metropolitan area. The I-270 Technology Corridor is also a major research and development center in the State of Maryland. Nevertheless, no significant manufacturing plants, such as chemical plants or refineries, are located near the reactor, and mining and quarrying operations are limited to those associated with constructing new office buildings. A natural gas pipeline lies 3.2 km (2 mi) south of the reactor, and a liquid petroleum/gas pipeline is located 1.6 km (1 mi) north.

Three arterial and collector roads abut the NIST campus boundaries: West Diamond Avenue forms the northern campus boundary; Quince Orchard Road the northwest boundary; and Muddy Branch Road the southeast boundary. The arterials and collectors serve the Gaithersburg area, providing truck access. Parallel to the northeast boundary of the NIST campus is a CSX rail line (CSX Transportation Corporation). At its closest point to the reactor, it is approximately 2 km (1.25 mi) away from the NIST boundary. This rail line carries goods and commuters through the region, providing service to the Maryland Rail Commuter (MARC) train in northern Montgomery County, Frederick County, and other points north for commuters traveling to Washington, D.C. The nearest MARC station is the Gaithersburg Station, 3 km (1.75 mi) away; the nearest Metro Station into the Washington, D.C., area is the Shady Grove Station, 5 km (3 mi) away.

Three commercial airports serve the region: Dulles International Airport (IAD) in northern Virginia is 29 km (18 mi) from the reactor, Ronald Reagan Washington National Airport (DCA) in Virginia just across the Potomac River from Washington, D.C. is 40 km (25 mi) away, and Baltimore-Washington International Thurgood Marshall Airport (BWI) near Baltimore, Maryland, is 47 km (29 mi) away. Andrews Air Force Base, the nearest military airbase, is approximately 52 km (32.5 mi) away. No normal air routes, holding patterns, or approach patterns associated with these airports cross the airspace above the NIST campus.

The Montgomery Airpark , a general aviation airport, is approximately 7 km (4.5 mi) northeast of the reactor and it lies 140°/320° relative to magnetic north; that is, it is nearly perpendicular to the line between the reactor and the airfield. Approximately 140,000 annual take-offs and landings occur at this field, with the typical air traffic consisting of small local aircraft, news aircraft, and an occasional military helicopter. The National Transportation Safety Board database (covering 1962 to the present) revealed 6 fatal air accidents and 18 non-fatal accidents in the Gaithersburg area. All but one of these accidents involved either airplanes or helicopters (one involved a balloon), and all were within 3 km (2 mi) of the Airpark. Small airplanes using the Airpark pose minimal risk to safe operation of the reactor.

Although there are a few recreational lakes within the area, the nearest major waterway is the Potomac River that forms the border between Maryland and Virginia. Its nearest point is 10 km (6.2 mi) from the reactor.

As described in the preceding sections, the NBSR is located in an urban setting with certain normal risks associated with transporting goods and materials on nearby highways and rail lines. These risks are regulated by several agencies, primarily by the U.S. Department of Transportation, to ensure safety. Also, the NIST campus serves as a buffer separating these transportation corridors from the reactor. The NIST campus also acts as a buffer between the NBSR and the surrounding community. This provides operators with greater control over the immediate area should there be an accident at the reactor.

2.1.2 Description of Reactor Complex

The NBSR is a heavy-water-moderated and -cooled, enriched-fuel, tank-type reactor designed to operate at 20-MWt (megawatts thermal power). It is a custom-designed variation of the Argonne CP-5 class reactor. The NBSR uses U_3O_8 aluminum dispersion fuel enriched to 93 percent. The fuel is aluminum-clad, materials-testing-reactor (MTR), plate-type fuel. The core is immersed in heavy water (D_2O) to slow the fast-moving neutrons that sustain the nuclear fission reactor, to dissipate heat created by the reaction, and to function as the first stage of shielding. Heavy water also allows high neutron fluxes that would not be otherwise achievable in a facility the size of the NBSR. This type of reactor (using MTR fuel and heavy water coolant) is similar to those used at a number of other government research facilities.

The primary coolant is also heavy water, which is circulated through a closed aluminum and stainless steel system. The heavy water is pumped through plate-type heat exchangers, where heat is transferred to a secondary cooling system before returning it to the core. The secondary system consists of plate-type heat exchangers and a plume suppression cooling tower that contains about 500,000 L (132,000 gal) of light water (H_2O). Heat in the secondary system is transferred to the atmosphere by evaporation of water from the cooling tower, which is located outside the confinement building.

The design of the NBSR includes many inherent passive safety features. The prompt neutron lifetime is relatively long as a result of heavy water moderation. The reactivity coefficients of void and temperature are negative. The reactor operates in a low-temperature, unpressurized condition and does not have a large stored energy content. Two inner structures within the reactor vessel retain heavy water in the event of a loss of water from the reactor core. In the event of a loss of cooling water, one of these structures immediately supplies emergency coolant flow to the fuel elements without any operator intervention, while the other maintains water around the lower half of the core. An overhead reserve tank can supply heavy water for emergency cooling either to the top or to the bottom of the elements for extended periods of time.

The NIST laboratory complex includes the NBSR confinement building, which is constructed of reinforced concrete and situated partially below grade. The complex includes nuclear-science-related research and other reactor support functions. The confinement building has an independent ventilation control system, and is capable of operating in isolation mode or dilution mode to exhaust air to the atmosphere through a 30-m (100-ft) stack.

2.1.3 Experimental Facilities

The NBSR is used for research, the majority of which uses neutrons to study material constituents, processes, and structure. The reactor design was chosen because of its thermalized (i.e., low energy) neutron spectrum, its high neutron flux, its flexibility for research, and its inherent safety. The high neutron fluxes generated by the NBSR are used in five principal ways:

1. to characterize the structure and dynamics of materials critical to the U.S. economy
2. to image large structures, and to study nuclear and neutron physics
3. to develop material and radiation standards
4. to generate radioisotopes for activation analysis and tracer production
5. to study the effects of radiation on materials through in-core irradiation.

Experimental facilities supporting these activities are described in the following sections.

The NBSR has a wide range of research capabilities and a large number of experimental beam lines. The liquid hydrogen cold source provides cold neutrons (i.e., neutrons slowed to speeds of 1000 m/s or less) directly to experiments in the confinement building, and through a network of seven neutron guides, to experiments located in the Cold Neutron Guide Hall. Beam tubes provide thermal neutrons for experiments located within the confinement area immediately adjacent to the reactor. A pneumatic "rabbit" system provides researchers with the ability to automatically inject samples into the core region of the reactor, while vertical thimbles provide for manual sample loading.

Eleven insertion positions are available for experiments within the core structure itself, and seven positions are available in the reflector. Nine beam tubes are arranged in a radial pattern within the central plane of the core and "see" the neutron flux in the unfueled gap region. Two beam tubes run completely through the reactor on either side of the core just below the radial tubes. The reactor includes a large experimental thimble within which a low-temperature, liquid-hydrogen moderator or cold source is installed. This moderator increases the intensity of cold neutrons available to the beams from this neutron source. Seven neutron guide tubes, which transport cold neutron beams with losses of less than 1 percent per meter into an adjacent neutron experimental building or neutron guide hall, and one beam port, which does not go to the Guide Hall, are served by this source. Five pneumatic tubes comprise the rabbit system that operates using pressurized carbon dioxide (CO_2). This system allows the rapid insertion and removal of small samples into various parts of the core, reflector, and thermal column. A large volume of well-thermalized neutrons is also available in the graphite thermal column.

2.1.3.1 Neutron Beams

The cross-sectional area of the neutron beams at the NBSR typically have ranges from a few mm^2 to 200 cm^2. Beams associated with an in-beam dose rate in excess of 1 mSv/hr (100 mrem/hr) and are accessible (i.e., have an open path in excess of 30 cm) are designated as High Radiation Areas. A characteristic of neutron beams is the radiation field outside of the beam is typically less than 0.05 mSv/hr (5 mrem/hr). Occasionally, experimental samples or equipment, such as collimators or filters, can result in Radiation Area or possibly High Radiation Area conditions near the beams. These areas are controlled as required by Title 10 of the Code of Federal Regulations (CFR) Part 20, Sections 1601 and 1902. Non-beam-related and short-term experiments are shielded and controlled to keep personnel exposures "as low as reasonably achievable" (ALARA).

2.1.3.2 Thermal Column Facility

The Thermal Column Facility provides highly thermalized neutron beams and is typically controlled as a High Radiation Area (per 10 CFR 20.1601). The facility is used to perform experiments requiring large cross-section exposures involving irregular exposure geometries or full-field exposure geometries.

2.1.3.3 Pneumatic System and In-Core Exposure Facilities

Experiments using the pneumatic system and in-core exposure facilities are highly variable, frequently producing multi-curie activity sources. ALARA concerns are addressed by shielding the source and by allowing sufficient decay time prior to direct manipulation, processing, or analysis. Technical review and administrative authorization processes are used to control these facilities' activities, usage, disposal, and potential personnel exposures.

2.1.3.4 Cold Neutron Experiments

The guides for cold neutron experiments are fully shielded to the point of neutron beam extraction where possible. At the entry wall to the Guide Hall, the unshielded dose rate from a typical guide is 3 mSv/hr (300 mrem/hr) (neutron) and 1 mSv/hr (100 mrem/hr) (gamma) at 1 m (3.3 ft) from the guide. All seven guides in the Guide Hall have primary shutters. These shutters are key-controlled and have status indicators (opened or closed). With the shutter closed, the design allows unrestricted access for disassembly and work on experiments associated with a particular guide.

2.1.4 Radioactive Waste Management Systems and Effluent Control Systems

NIST has a structured radiation protection program that supports all aspects of NBSR operations. The health physics staff is equipped with radiation detection equipment to determine, control, and document all occupational radiation exposures. An environmental monitoring program is in place to determine if potential radiation exposures to members of the public in unrestricted areas surrounding the reactor remain within regulatory standards and guidelines.

The overall radioactive waste management and effluent control programs for the NBSR are described in this section. NIST has established policies that employ the ALARA concept in all operations at the NBSR, and operations at the NBSR and experimental facilities are conducted to minimize radioactive effluents and waste production consistent with ALARA objectives.

Reactor and the Environment

2.1.4.1 Radiation Sources

Sources of radiation monitored and controlled by the radiation protection and radioactive waste management programs are described in this section.

Radiation sources at the NBSR can be classified into four general classes:

- Calibration and check sources
- Startup sources and other sources used for instrumentation and nuclear support functions
- Gaseous, liquid, and solid radiation sources from reactor operations
- Radiation sources produced within the experimental facilities.

Sources of radioactivity that may be found in various reactor and support systems are listed in Table 2-6.

Table 2-6. NBSR Systems and Radiation Sources

NBSR System	Major Sources of Radioactivity	Minor Sources of Radioactivity
Primary coolant	H-3, N-16, Co-60	Ar-41, Na-24, Mn-54, Mn-56, Cr-51, Sb-122, Sb-124
Primary pipe (internal contamination)	Co-60, H-3	Cr-51, Zn-65
Helium sweep	Ar-41	Kr-85m, Kr-87, Kr-88, Xe-131m, Xe-133, Xe-135, Xe-135m, Xe-138, Cs-138
Thermal shield cooling system	Cu-66, Cu-64, Ag-110m, Zn-65	N-16
Reactor shield plug/refueling plug	-----------	Al and steel activation products, C-14
Air	Ar-41, H-3	Br-82, Cl-38, Cs-138
CO_2 sweep gas	Ar-41	Br-82, Cl-38, S-35
Storage pool	H-3	Aluminum activation products from fuel cutting
Fuel pieces (6061 aluminum, stainless steel)	Fe-55, Co-60, Zn-65	Ni-63, Mn-54
Resin beds	Co-60, Zn-65	---------------
Neutron guides	Zn-65	Co-58, Ni-59
Pneumatic system	Co-60, Ag-110m, Zn-65	---------------

2.1.4.2 Liquid Waste Processing Systems and Effluent Controls

The dominant radionuclides in liquid effluents at the NBSR are tritium (H-3) and N-16. Other minor liquid sources are also discussed in subsequent sections.

Reactor Primary Coolant

The NBSR primary coolant consists of high-purity heavy water. The radionuclides in liquid effluents at the NBSR are primarily tritium and N-16.

The following reactions produce most of the radioactive materials in the primary coolant:

- Tritium, a low-energy beta-emitter, produced via H-2(n,γ)H-3
- N-16, a high-energy beta- and gamma-emitter, produced via O-16(n,p)N-16
- Na-24, a high-energy beta- and gamma-emitter, produced via Al-27(n,α)Na-24
- Al-28, a high-energy beta- and gamma-emitter, produced via Al-27(n,γ)Al-28
- Co-60, a low-energy beta- and high-energy gamma-emitter, produced via Co-59(n,γ)Co-60
- Cr-51, a low-energy gamma-emitter, produced via Cr-50(n,γ)Cr-51.

Other radionuclides in the primary coolant that contribute minor portions to the total liquid radiation source include Zn-65, Mn-56, Tc-99m, and Sb-122, which are associated with suspended corrosion products activated by neutrons.

Tritium could potentially be a significant source of exposure from airborne contamination because of evaporation of tritiated heavy water. Either inhalation or exposure by direct contact through skin absorption could result in significant exposures. Therefore, any work involving potential exposure by these mechanisms requires control measures, such as containment, eye protection, gloves, and protective clothing, to minimize and prevent such an occurrence. Individuals who perform this work are required to undergo periodic tritium bioassays. Other radionuclides are present at such low concentrations they have minimal potential for intake via inhalation or skin absorption.

N-16 is the greatest operational source of external radiation exposure from the primary piping system. N-16 has a short half-life (7 seconds), so exposure from this source diminishes very rapidly after the reactor is shut down. At the NBSR, the Process Room and the Monitoring Room are areas where a potential for exposure from N-16 exists.

Na-24 is present in the primary coolant at concentrations on the order of 0.1 mCi/L. It represents a transient source of external exposure in the process room. Because of its short half-life (15 hours), work in the process room is limited for the first day following shutdown as an ALARA measure.

Other than tritium, Cr-51 is the highest activity and longest-lived (half-life 27.7 days) primary system contaminant. Because it is a low-energy gamma emitter, Cr-51 is almost completely self-shielded by the primary system components. The activity is dominated by Cr-51 in primary components immediately following removal from the reactor.

Because Cr-51 decays relatively quickly, Zn-65 and Co-60 become the dominant sources of residual contamination after several months. Localized external contamination occurs at valves, heat exchangers, filters, and resin beds, and ranges from a few hundredths of a mSv/hr (few mrem/hr) to 0.5 Sv/hr (50 rem/hr). Control of personnel exposure is accomplished through shielding and posting of areas. Components with higher dose rates, such as primary coolant filters and resin beds, are shielded to reduce the radiation levels to less than 0.05 mSv/hr (5 mrem/hr), and exposures from other areas are controlled through local posting. The general area dose rates in the process room are routinely surveyed because of the cumulative effect of the longer-lived internal contaminants. Process room survey data are made available for work planning.

Other potential, but unlikely sources of radionuclides from liquid sources include the reactor secondary coolant system, the thermal column D_2O tank coolant system, the thermal shield cooling system, and the fuel storage pool.

Liquid Waste

The liquid waste collection facility consists of a 3785-L (1000-gal) tank, two 18,900-L (5000-gal) tanks, various filters, and related pumps and valves. Water is collected, sampled, and analyzed for its radioactive constituents and then filtered to meet 10 CFR 20.2003 solubility requirements before being released to the sanitary sewer. Credit is taken for the daily NIST site release volume of approximately 984,100 L (260,000 gal) to meet the 10 CFR 20.2003 concentration limits.

If unanticipated quantities of radioactive material are accumulated in the system, the contaminated water can either be circulated through filters or resin beds to reduce the radionuclide concentration, transferred to containers for offsite processing at a NRC licensed facility, or stored to allow radioactive decay to reduce the level of activity. When practicable, a general ALARA operating practice at the NBSR is to collect any higher activity liquid wastes at the source and to process and dispose of that waste separately.

2.1.4.3 Gaseous Waste Processing Systems and Effluent Controls

Three gaseous waste streams associated with the reactor facility include the normal air, irradiated air, and process room ventilation systems. Processes that might generate airborne particulate or gaseous contamination are vented through one of these systems. Gases in these systems are passed through high efficiency particulate air (HEPA) filters prior to release via the stack. For an upset or abnormal operating condition, these ventilation systems can be operated in recirculation mode, at which time a standby charcoal filter becomes operational. Monitoring systems in the stack and in the building ventilation use both installed and periodic sampling. This practice provides redundant methods for assessing and controlling both occupational and public exposure.

Ar-41 Sources

Ar-40 is a natural constituent of air, present at about 0.93 percent. Any air volume that is exposed to neutrons will contain Ar-41 produced by the Ar-40(n,γ)Ar-41 reaction. Ar-41 is a strong beta- and gamma-emitter with a half-life of 110 minutes. At the NBSR, engineering and procedural measures have been established to minimize Ar-41 production.

Tritium

The tritium produced by the heavy water moderator/coolant of the reactor yields a primary coolant tritium concentration of 1.1×10^{10} Bq/L/yr (0.3 Ci/L/yr). As an ALARA measure, NIST replaces the heavy water at intervals designed to limit exposure to tritium. All used heavy water is stored onsite until transfer to authorized processors for recycling. With a maximum tritium production concentration of 18.5×10^{10} Bq/L (5 Ci/L), the exposures discussed in this section for the NBSR would increase by no more than a factor of five.

During normal operations, the primary release pathway for tritium results from helium leakage into the ventilation system. Helium can become saturated with heavy water vapor when used as a cover gas to minimize air intrusion into the primary cooling system. Activation of heavy water in the helium cover gas produces tritium. Secondary pathways include refueling or any maintenance activity that exposes heavy water to the air. Conditions involving an abnormal loss of coolant, such as a seal failure or a primary coolant boundary failure, would be identified by monitoring and leak detection systems. The airborne tritium monitoring system at the NBSR is capable of detecting tritium concentrations that can occur by water evaporation following a few milliliters of leakage.

Fission Products

Noble gas fission products, including gaseous xenon, krypton, and Cs-138 (a decay product of Xe-138), can be detected in the helium sweep system over the primary coolant. Based on the typical make-up rate for the helium system, less than 3.7×10^9 Bq (0.1 Ci) of those radionuclides are released annually, resulting in release concentrations so low that they represent a negligible contribution to the total gaseous emissions.

Air Monitoring

Conditions requiring airborne radioactivity monitoring under 10 CFR 20.1502(b) are rarely present at the NBSR. Two primary airborne radionuclides are detectable at the NBSR: Ar-41 and tritium. Area radiation monitors are used to control personnel radiation exposures to Ar-41, and Cary ion chambers or gas Marinelli chambers detect airborne activity concentrations with a sensitivity greater than 0.1 derived air concentration (DAC). For tritium, an installed gas-flow ion chamber system samples representative areas of the NBSR building and its ventilation system; it can detect tritium concentrations of 0.1 DAC and is sensitive to Ar-41. A cold trap also samples for tritium, with samples analyzed using liquid scintillation at a sensitivity greater than 10^{-6} DAC.

In addition to the Ar-41 and tritium monitoring, continuous air monitors (CAMs) are available for airborne particulate and iodine monitoring on an as-needed basis. For instance, one CAM is located in the spent fuel storage pool area. Filter and charcoal cartridge samplers may also be used for iodine and particulate sampling.

Effluent Monitors

Airborne effluent at NBSR is monitored for Ar-41 using a G-M (Geiger-Mueller) detector located in the stack. This system is calibrated by comparison to a grab sample that is analyzed in the radioanalysis laboratory. The nominal monitor sensitivity is 1.4×10^{8} μCi/m.

Tritium in the NBSR stack effluent is continuously monitored by the building tritium monitoring system. Monthly grab samples from the stack are also collected and analyzed for verification. More frequent sampling or additional continuous monitoring is implemented when unusual or non-routine activities involving the potential for additional tritium releases are performed. Effluent sampling can also be performed with a particulate filter and charcoal cartridge, which are analyzed on an as-required basis.

Environmental Monitors

Environmental (ambient) monitoring is accomplished in several ways. Ambient gamma-monitoring is conducted with thermoluminescent dosimeters (TLDs), by a pressurized tissue-equivalent ion chamber system with sensitivity of 0.1 μrad/hr, by environmental G-M monitors with data logging, and by a gain-stabilized sodium iodide system (with a sensitivity of 0.01 μrad/hr for Ar-41) for monitoring Ar-41 or other specific gamma-emitters.

2.1.4.4 Solid Waste Processing

Solid radioactive waste is any contaminated item having no further usefulness, and for which further decontamination is not practicable. Radioactive wastes are segregated from non-radioactive wastes based on knowledge of where the material was used or from which system it originated. Items that are exposed to neutrons or to sources of contamination are considered potentially radioactive, including irradiated hardware from experiments or items that came in contact with primary reactor coolant. On occasion, process knowledge suggests an item can be decontaminated. If an item is successfully decontaminated, as determined by a radiation survey and contamination check, it may be released for unrestricted use or disposal.

Solid Radiation Sources

Reactor operations include solid sources of radiation at the NBSR, ranging from items having very low specific activity (e.g., used rubber gloves from handling potentially contaminated materials) to intermediate activity (e.g., activated foils from experiments), and high activity (e.g., spent fuel from the reactor).

Fuel Elements

All operations involving movement of irradiated reactor fuel elements are performed underwater to provide shielding. An underwater saw is used to separate non-fuel portions of the spent fuel elements from the fueled portions of the elements, and they are disposed of separately. The dominant radionuclides are Fe-55, Co-60, and Zn-65. Shielding is used to reduce personnel exposure during spent fuel handling operations.

The fission product inventory for one NBSR fuel element includes radionuclides with a total activity of 1.5×10^{16} Bq (3.97×10^5 Ci). Personnel protection is needed primarily to reduce dose rates from the fuel elements, and all fuel transfers are performed within a shielded pathway. The room through which the elements are transferred is controlled as a Very High Radiation Area during these transfers per 10 CFR 20.1602 requirements. All fuel-handling in the storage pool is monitored with area monitors or survey instruments to determine that shielding is adequate.

New NBSR fuel elements nominally contain 350 g (0.77 lb) of U-235 and are surveyed for external radiation levels and surface contamination when they are received. Each element undergoes a quality assurance evaluation before being inserted in the reactor. Dose to operators when handling new unirradiated fuel is minimal because there are no fission or activation products present.

Other Radioactive Solids

Other radioactive solids that contribute to personnel dose and waste volume include the following:

- Reactor shims
- Reactor primary resins, replaced once every 10 to 20 years
- Reactor primary filters, replaced as needed, usually once or twice per year

- Filters and resins from other systems
- Shielding plugs and related neutron beam shields
- Experiments or experimental components removed from high neutron flux locations
- Activated experiment samples and
- Miscellaneous contaminated materials, such as laboratory waste.

The radioactive material content of these items ranges from barely detectable levels in the bulk of the waste volume, to curie-quantity material for specific items such as resins. The primary contributor to personnel external dose rate is the Co-60 in the activated metals, resins, and much of the other waste. Material contaminated with Co-60 is stored in restricted areas where access and area dose rates are controlled. Local shielding is used as necessary to limit spaces to less than Radiation Area conditions. Two storage areas are maintained as restricted areas. The concrete shield cave facility in the G-Wing is used to store shielded casks. Other items, such as bulky items with low-level activation (e.g., experiment shields and components), may be stored in Building 418 adjacent to the reactor building.

Solid Radioactive Waste Characterization and Disposition

Solid radioactive waste is characterized by direct assay, which involves sampling and direct gamma spectroscopy, as well as by process knowledge.

Solid radioactive waste is accumulated at the point of production and collected consistent with keeping exposures ALARA. All accumulation containers are appropriately labeled. Collected low-level waste is typically transferred to the H-Wing. Records of the origin of the waste and its radiological contents are kept in preparation for packaging and shipment. Other waste requiring special handling or containing high levels of radioactivity, such as primary filters and large neutron beam shields, is stored at other locations.

Systems, components, and experiments are designed to minimize the production of mixed waste (which contains both chemically hazardous and radioactive constituents) to the maximum extent practicable. Any such waste (e.g., lead or cadmium) that has been exposed to neutrons is segregated and stored until disposal at an authorized facility is arranged.

All radioactive waste is disposed of in accordance with 10 CFR Part 20, Subpart K. Solid waste is transferred to organizations specifically authorized or licensed to receive the material, such as permitted commercial treatment and disposal facilities or the U.S. Department of Energy (DOE). Materials designated as radioactive waste are transferred to the H-Wing of the NBSR for characterization, packaging, and preparation for transfer to authorized recipients. Annual radioactive waste volumes during 2001 to 2005 ranged from 12 to 16 m^3 (440 to 574 ft^3). During that period, the total radioactivity in waste shipments designated Class A under 10 CFR Part 20 was less than 5.6×10^{10} Bq (1.5 Ci). Two shipments of Class C waste during the same period contained a total of about 5.2×10^{13} Bq (1400 Ci) of radioactive material. Larger quantities of radioactive waste may be generated in years when unfueled element shipments occur, or when major facility modifications are performed. Based on past experience, these events occur on the order of once every 5 or more years. No radioactive waste designated as Greater than Class C or transuranic waste has been generated at the Center for Neutron Research, nor is such waste anticipated in the future.

All solid radioactive waste is disposed of by transfer to either licensed disposal sites or processing facilities. It is transported as required by 10 CFR Parts 61 and 71 and by the applicable licenses issued by states to the receiving facilities. Detailed radioactive waste characterization documents and manifests are prepared and retained in accordance with 10 CFR 20.2006.

Reactor and laboratory operations generate small quantities of mixed low-level waste (MLLW), which contains both radioactive and chemically hazardous components. Solid MLLW consists mainly of activated cadmium and lead experimental components or shielding and is generated at the rate of about 0.06 m³ (2 ft³) per year. Removal of reactivity control blades from the reactor accounts for an additional 0.06 m³ (2 ft³) of MLLW about every 8 years. The control blades are stored for 7 years to allow radioactive decay so they can be disposed of as Class A waste. Liquid MLLW consists of contaminated cleaning solvents or organic assay solutions generated at the rate of about two 55-gal drums (about 420 L) per year. MLLW is treated as required prior to disposal at facilities specifically permitted for such waste.

Some structural components of the reactor and neutron beam ports also contain lead and will require disposal as MLLW upon reactor decommissioning. Those components include the reactor thermal shield, which consists of approximately 114,000 kg (250,000 lb) of lead bonded to carbon steel. An additional 4.3 m³ (150 ft³) of MLLW consists of neutron beam-port shutters that contain lead incorporated into stainless steel alloys. Those components have been placed into long-term storage until decommissioning.

Spent fuel is generated at a rate of approximately 28 elements per year and is stored onsite in a storage pool until it can be cut into sections and shipped. Each element contains two 34.37 x 8.55 x 7.62 cm (13.5 x 3.4 x 3.0 in.) fueled sections, which are shipped to DOE's Savannah River Site. Approximately 252 sections with a total volume of about 0.6 m³ (20 ft³) are shipped to the Savannah River Site every 4½ years. The unfueled sections of the fuel elements are segregated and disposed of as low-level radioactive waste.

Solid Waste Minimization

Because the costs of solid radioactive waste disposal are high, materials with low activation potential are used wherever practical to minimize the production of radioactive waste. At the NBSR, experiments are designed to be reusable and to minimize the generation of radioactive material by neutron activation. Radioactive contamination of materials used in experiments and processes is also minimized to the extent practicable.

Components are disassembled and segregated where possible to minimize quantities of radioactive waste. To the extent practicable, a commercial, HEPA-filtered compactor is used to reduce the volume of compressible materials such as laboratory paper waste and contaminated gloves.

Long-Term Storage

The policy at the NBSR is to dispose of items identified as waste in a timely manner. There is a long-term storage area located in the G-Wing of Building 235 to accommodate radioactive materials that require storage prior to disposal or for potential reuse. The facility contains 33 shielded concrete cavities, each about 3 m (10 ft) deep and varying in diameter. The shielded facility is used to store items that could produce a significant exposure to workers, but which have potential future use. It is also used to store some higher-activity items to allow radioactive decay prior to disposal.

2.1.5 Nonradioactive Waste Systems

NIST does not dispose of any non-hazardous solid waste onsite. Tree limbs, shrubs, and other organic matter are chipped, stockpiled, and reused as mulch. During 2005, the NIST site recycled approximately 760 tonnes (840 tons) of waste materials consisting of scrap metal, computers, electronics, paper products, cans, glass, plastic, fluorescent light bulbs, lead-acid batteries, waste oil, mercury, and other chemicals. The remaining non-hazardous solid waste generated at the NIST site, estimated at about 45 tonnes (50 tons) per year, is sent to Montgomery County solid waste processing facilities.

2.2 Interaction of the Reactor with the Environment

The siting requirements contained in 10 CFR Part 100 apply to applications for site approval for the purpose of operating stationary nuclear power reactors as well as testing reactors. The site evaluation criteria for the NBSR at the NIST Center for Neutron Research within the NIST campus are defined in 10 CFR Part 100, Subpart A, "Evaluation Factors for Stationary Power Reactor Site Applications Before January 10, 1997, and for Testing Reactors."

The following sections provide general descriptions of the environment near NIST as background information. They also provide detailed descriptions where needed to support the analysis of potential environmental impacts of operation during the renewal term, as discussed in Chapter 3. Section 2.2.11 describes possible impacts associated with other Federal project activities. The discussions presented in this chapter are based on reviews of the most recent site-related information, several past reports, and information published since the last application for license renewal and power upgrade that would have an impact on site safety.

2.2.1 Land Use

The NBSR is located on the NIST campus in an unincorporated portion of Montgomery County, Maryland. The campus is approximately 32 km (20 mi) northwest of Washington, D.C. The NBSR is part of the NIST Center for Neutron Research.

The NIST campus encompasses 234.5 ha (579.5 ac). The Center for Neutron Research reactor-laboratory complex is located on Center Drive in the southern portion of the NIST campus. The NIST campus is located between several major roads. The northeast boundary of the campus abuts I-270, a major commuter artery connecting communities in northern Montgomery County, Frederick County, and other points north to the employment areas in the Washington, D.C., metropolitan area. West Diamond Avenue forms the northern boundary of NIST, with Quince Orchard Road as the northwest boundary, and Muddy Branch Road as the southeast boundary. The closest railway parallels the northeast boundary of the NIST campus at a distance of approximately 2 km (1.25 mi) from the NBSR at its closest point. This line carries goods and commuters through the region. The nearest waterway to the NIST campus is the Potomac River, which forms the border between Maryland and Virginia. Its nearest point is approximately 10.3 km (6.4 mi) from the NBSR.

Montgomery County is not within Maryland's coastal zone for purposes of the Coastal Zone Management Act (MDNR 2002).

2.2.2 Water Use

The NIST reactor uses from 568,000 to 662,000 L/d (150,000 to 175,000 gal/d) of water from the Washington Suburban Sanitary Commission's (WSSC) water supply system. The primary consumptive use of this water is associated with the cooling towers, which provide secondary cooling for the reactor. The average loss of 376,000 L/d (100,000 gal/d) from evaporation and drift from the reactor's cooling towers represents less than 0.1 percent of the WSSC's average capacity. The sources of WSSC's water are the Potomac and Patuxent Rivers. About 24 percent of the water withdrawn from the WSSC system is returned as blowdown to the WSSC sanitary system.

2.2.3 Water Quality

The NIST reactor discharges non-radiological liquid effluents to the WSSC sanitary sewer system. The majority of the effluent is blowdown from the cooling towers. The blowdown contains zinc from corrosion prevention measures and elevated dissolved solids from evaporative concentration of existing dissolved solids in the makeup water. NIST operates under the WSSC Discharge Authorization Permit (05813). The permit was issued by the State of Maryland on June 1, 2004, and is scheduled to expire on May 31, 2008.

2.2.4 Meteorology and Air Quality

The NIST site is located on the Piedmont Plateau of Maryland, a transitional region between the Blue Ridge Mountains to the west and the Atlantic Coastal Plain to the east. Because of its mid-latitude location and elevation of approximately 128 m (420 ft), the site's climate is classified as continental, with four distinct seasons.

The climatology of the NIST site can be described using archived data from two nearby National Weather Service (NWS) observing stations: Dulles International Airport (IAD) (NCDC 2005a) and Ronald Reagan Washington National Airport (DCA) (NCDC 2005b). Although an AWS Convergence Technologies Inc. WeatherNet Weather Station has been installed on the roof of the NIST confinement building since 2002, there is not a sufficient period of data to develop a complete climatology of the site from this data set alone (NIST 2004a).

Normal daily maximum temperatures for IAD range from a high of 30.8°C (87.4°F) in July to a low of 5.2°C (41.4°F) in January. Normal daily minimum temperatures range from 17.8°C (64.0°F) in July to -5.6°C (21.9°F) in January. At DCA, average temperatures are somewhat greater, especially normal daily minimum temperatures, which is a result of the station's location in relation to the surrounding city, resulting in a phenomenon called the urban heat island effect.

Precipitation is distributed evenly throughout the year, with annual liquid precipitation amounts averaging 106.17 cm (41.80 in.) at IAD and 99.95 cm (39.35 in.) at DCA. Spring and summertime precipitation is generally from thunderstorms, whereas the bulk of autumn and winter precipitation is from large-scale weather systems moving through the region. Occasionally during the late summer and autumn months, tropical storm remnants can affect the area, bringing widespread and significant precipitation events. Indeed, the greatest 24-hour precipitation amounts of 30.18 cm (11.88 in.) at IAD and 18.26 cm (7.19 in.) at DCA were from Hurricane Agnes as it passed east of the region as a tropical storm on June 21-22, 1972 (NCDC 2005a, b).

Annual average snowfall amounts for the area range from 53.85 cm (21.2 in.) at IAD to 38.61 cm (15.2 in.) at DCA, where it tends to be slightly warmer. January is usually the snowiest month, with two days averaging above 2.54 cm (1 in.) of snowfall. Heavy snowfall events, though rare, do occur and can bring some 50.8 cm (20 in.) of snow in a 24-hour period to the region.

Thunderstorms occur approximately 30 days out of the year, with the majority of the thunderstorms occurring during the months of May through August (NCDC 2005a, b). On occasion, these storms are severe, with gusty winds and hail the primary threat. On average, there are 1.1 high wind events and 2.1 hail events per year in Montgomery County (NIST 2004a).

Tornado climatology statistics from 1950 through 2003 also show that 83 tornadoes have occurred within a 1° box that includes the NIST site (Ramsdell 2005). Of these, only 13 tornadoes had intensities of F2 or F3 (winds between 113 and 206 mph) on the Fujita intensity scale, and no reported tornadoes had intensities of F4 or greater. The probability of a tornado striking the site is expected to be 8.3×10^{-5} per year (Ramsdell 2005).

The average wind direction for the region is bimodal, with southerly winds dominating during the summer and northwesterly winds from mid-autumn through early spring. The change in direction results from different influencing features: the Bermuda High during the summer months and large-scale weather systems and associated fronts during the winter months. Wintertime winds average around 3.8 m/s (8.5 mph), whereas summertime winds tend to be weaker, averaging around 2.9 m/s (6.5 mph). Occasionally, wind gusts can reach 22.4 to 26.8 m/s (50 to 60 mph) from passing fronts, thunderstorm outflow, or tropical storms (NCDC 2005a, b).

The NIST site is in Montgomery County, Maryland, which is part of the National Capital Interstate Air Quality Control Region (AQCR) (40 CFR 81.12). This AQCR also includes the District of Columbia, Prince Georges County in Maryland, and Arlington, Fairfax, Loudoun, and Prince William Counties in Virginia.

With respect to criteria pollutants regulated under the National Ambient Air Quality Standards (NAAQS), Montgomery County is designated as unclassifiable, in attainment, or better than the national standards for nitrogen dioxide, sulfur dioxide, carbon monoxide, and total suspended particulates (40 CFR 81.321). On March 25, 2003, this County was designated as in severe nonattainment to the 1-hour ozone standard and more recently (June 15, 2004) designated as in moderate nonattainment to the newly promulgated 8-hour ozone standard (40 CFR 81.321). In addition, Montgomery County is in nonattainment for fine particles, which are particles with a diameter of 2.5 µm or less (PM2.5) (40 CFR 81.321).

The Air Quality Index (AQI) is a national standard method for reporting daily air-pollution levels to the general public (40 CFR Part 58, Appendix G). The AQI is a composite index based on the criteria pollutants that are in the NAAQS. Depending on the value of the index, days are classified as Good, Moderate, Unhealthy for Sensitive Groups, Unhealthy, Very Unhealthy, and Hazardous. For the 5 years from 2001 through 2005 in which the AQI was calculated, 76 percent of the days for Montgomery County were classified as Good, 21.5 percent were classified as Moderate, 2.4 percent were classified as Unhealthy for Sensitive Groups, and 0.1 percent were classified as Unhealthy (U.S. EPA 2005).

Emergency power generators and other facilities and activities associated with the NIST site emit various pollutants, which are regulated under Title V Operating Permit 24-030-00323 by the Maryland Department of the Environment, Air Quality Permits Program, Air and Radiation Management Administration; the permit is scheduled to expire on April 30, 2008.

2.2.5 Aquatic Resources

The NBSR is located within the Seneca Creek/Anacostia River sub-watershed of the Middle Potomac-Catoctin watershed of the Potomac River. The major rivers in this watershed generally flow in a southerly direction and eventually drain into the Chesapeake Bay. Tributary A of the Muddy Branch is the closest natural water body to the NBSR and is approximately 305 m (1000 ft) west-northwest of the reactor building. This tributary flows through an onsite storm-water retention pond and continues into Lake Varuna before entering the Muddy Branch. There is another unnamed tributary (called Tributary B) to the Muddy Branch located about 580 m (1900 ft) southeast of the site. A topographic rise separates this tributary from the reactor site (NIST 2005). The Muddy Branch supports a warm-water fish community including Bluntnose minnow (*Pimephales notatus*), swallowtail shiner (*Notropis procne*), and redbreast sunfish (*Lepomis auritus*) (Mongomery County Department of Environmental Protection 2006). The Muddy Branch enters the Potomac River near Katie Island, approximately 10 km (6.2 mi) southwest of the NBSR (NIST 2005; U.S. NRC 1982).

Surface water and groundwater are not used as process water in either the primary or secondary coolant systems. Water lost through evaporation from the secondary coolant system is replenished by the WSSC via municipal water supply lines, and the blowdown from the cooling towers is discharged into the sanitary sewer system. Process water is not discharged to surface water or groundwater (NIST 2005).

Water samples are collected from streams and ponds from a minimum of four locations as part of the Environmental Monitoring Program. Samples are collected all year, depending on availability. These samples are analyzed for possible activation radionuclides and fission products as well as assayed for tritium (NIST 2005).

There are no Federally listed aquatic species under the Endangered Species Act that occur in Montgomery County (MDNR 2004).

2.2.6 Terrestrial Resources

The NIST campus is located on the Maryland Piedmont Plateau about 48 km (30 mi) southeast of the Blue Ridge Mountains (NIST 2005). Common species that occur on the campus include Canada geese (*Branta canadensis*) and white-tail deer (*Odocoileus virginianus*). To better manage its deer population, NIST has partnered for the past decade with the Humane Society of the United States in the use of an innovative scientific means of birth control for wildlife (Newman 2005).

The NBSR is located in the Center for Neutron Research facility on the southern part of the NIST campus. The portion of this facility directly under the NRC's license consists of several buildings or parts of buildings, storage areas, and the cooling towers. There is also a parking lot, small amount of lawn, and landscaped gardens (NIST 2005).

Grass and soil are routinely sampled as part of the Environmental Monitoring Program. Soil samples are collected during the non-growing season (October through March), and grass samples are collected during the normal growing season (April through September). The collected samples are analyzed for possible neutron activation nuclides and fission product nuclides (NIST 2005).

One Federally listed endangered terrestrial animal species, the small whorled pogonia (*Isotria medeoloides*), is known to occur in Montgomery County (MDNR 2004). Although there is suitable habitat for the small whorled pogonia on the NIST campus, there are no known records of this species occurring on the NIST campus (U.S. NRC 2007).

2.2.7 Radiological Impacts

NIST conducts a radiological environmental monitoring program (REMP) in the vicinity of the NBSR. Through this program, radiological impacts to the public and the environment are monitored, documented, and compared to the appropriate standards. The objectives of the REMP are as follows:

- Provide representative measurements of radiation and radioactive materials in the exposure pathways and of the radionuclides that have the highest potential for radiation exposures to members of the public.

- Supplement the radiological effluent monitoring program by verifying that the measurable concentrations of radioactive materials and levels of radiation are not higher than expected on the basis of the effluent measurements and modeling of the environmental exposure pathways.

Results of measurements of radiological releases and environmental monitoring are summarized in annual operations reports to the NRC (NIST 2002, 2003, 2004b, 2005, 2006). The limits for all radiological releases are specified in the NBSR technical specifications (NIST 2004c), and these limits are designed to meet Federal standards and requirements. The REMP includes monitoring of the atmospheric environment (airborne radioiodine, gross beta, and gamma), the terrestrial environment (crops, soil, and milk), and direct radiation.

2.2.7.1 Environmental Monitoring

The NBSR Environmental Monitoring Program is designed to verify that radiation doses to the public remain within the limits set out in 10 CFR 20.1301. Through this program, staff at the NIST Center for Neutron Research perform effluent sampling and monitoring, environmental surveys, and liquid waste release monitoring. Because operational releases normally represent a negligible fraction of the regulatory limits, the real-time monitoring instruments displayed in the reactor control room are capable of recognizing a potential elevated release. Reviews of the recorded release data are also performed quarterly. Estimates of dose to members of the public are based on measured emissions and are determined by computational models. The U.S. Environmental Protection Agency (EPA) COMPLY code (U.S. EPA 1989) and other models are used to estimate doses.

Environmental surveys include radiation surveys, sampling of grass and soil, and sampling of water from local streams and ponds. Thermoluminescent dosimeters are used to detect direct radiation at the NIST site boundary. The collected samples are analyzed for possible activation and fission-product radionuclides. Water samples are also assayed for tritium. Samples of water, soil, and grass are collected and analyzed at least quarterly from a minimum of four locations for each type. Soil samples are collected during the non-growing season (October through March), and grass samples are collected during the normal growing season (April through September). Environmental analysis of soils and grasses typically has a sensitivity of better than 1 pCi per sample; liquid scintillation analysis of water samples typically has a sensitivity better than 10 pCi/mL.

Review of historical data on releases and the resultant dose calculations indicated the doses to maximally exposed individuals in the vicinity of the NBSR site were a small fraction of the limits specified in the EPA environmental radiation standards 40 CFR Part 190 as required by 10 CFR 20.1301(e). Dose estimates are calculated for a hypothetical maximally exposed individual, based on monitored liquid and gaseous effluent release data, onsite meteorological data, and appropriate exposure pathways.

2.2.7.2 Impacts From Radiological Liquid Emissions

The maximum annual dose to a member of the public from liquid effluents was less than 0.01 mSv/yr (1 mrem/yr) based on effluent radionuclide concentrations and values in 10 CFR Part 20, Appendix B, Table 3. Tritium is the dominant radionuclide in liquid effluents at NBSR, with annual releases from 2001 to 2005 on the order of 9.6×10^{10} to 18.1×10^{10} Bq (2.6 to 4.9 Ci), which would comply with the limits specified in 10 CFR 20.2003(a)(2) and (3). The NIST records for annual releases of other prominent beta-gamma emitters include Co-60, Zn-65, and Ag-110m. Liquid releases to the sanitary sewer (under the NIST materials license SNM-362) constitute a small fraction of the total NBSR liquid radioactive effluent. The annual volume of radioactive effluent released is typically about 1,135,500 L (about 300,000 gal), which is diluted by the NIST site sanitary sewer volume of approximately 379-million L (100-million gal). The major contributor to the liquid waste volume consists of air-conditioning condensate from the confinement building, which has low-level tritium contamination from the building air.

2.2.7.3 Impacts From Radiological Air Emissions

The principal airborne sources of radioactivity associated with operation of the NBSR are Ar-41 and tritium. The only release path for air from the various confinement building ventilation systems is via the building stack exhaust, which has a nominal flow rate of 30,000 cfm (14.2 m^3/sec). Between 2001 and 2005, annual emissions of Ar-41 ranged from 2.96 x 10^{13} to 4.4 x 10^{13} Bq (800 to 1200 Ci), tritium ranged from 2.6 x 10^{13} to 5.2 x 10^{13} Bq (700 to 1400 Ci), and other radionuclides contributed less than 7.4 x 10^9 Bq/yr (0.2 Ci/yr) on average.

The NRC ALARA dose constraint for radionuclides released to the atmosphere is 0.1 mSv/yr (10 mrem/yr) to any member of the public (10 CFR 20.1101). The dose to a maximally exposed individual from all air pathways during the period from 2001 to 2005 was less than 0.01 mSv/yr (1 mrem/yr) to the whole body or any organ other than the thyroid. This represents less than 10 percent of the NRC public dose constraint for exposure via air pathways. The gaseous exposure pathways included inhalation, ingestion of milk and crops, and direct radiation from the airborne radioactive material. This analysis was performed with the EPA COMPLY computer code (U.S. EPA 1989) using local meteorological data. The dose was estimated for the closest resident in each sector, which constitutes conservative analytical boundary conditions. These doses are typical of the annual dose for operation of the NBSR, and they are expected to remain well below NRC and EPA limits during the license renewal term.

From a public dose perspective, tritium results in about one-tenth of the dose from Ar-41, assuming release of equal activities. Conducting operations in a way that minimizes Ar-41 production, even if that results in some increased heavy water loss and minor increases in tritium exposure, results in minimized collective dose because the increased occupational dose to the limited number of operational staff is more than offset by the reduced collective dose to the public. Therefore, ALARA efforts to reduce tritium losses, particularly through ventilation system modifications, must consider possible related increases in Ar-41 emissions.

2.2.7.4 Dose to Workers

Ar-41 is produced at the NBSR primarily by neutron activation of air in the cavity around the reactor vessel. A secondary source is associated with experiments, which contribute less than 0.1 percent to the total. The external exposure rate from Ar-41 is minimal because the concentrations of Ar-41 in the building are less than 1 DAC and the building volume represents a small fraction of a "semi-infinite" cloud; actual dose rates to a person in the building from a uniform cloud at 1 DAC would be less than 0.2 mrem/hr. Personnel dose rates from typical Ar-41 levels measured inside the confinement building have been less than 4 x 10^{-5} mSv/hr (0.004 mrem/hr). Combined with typical occupancy times and reactor operating frequency, this would result in personnel exposure less than 0.02 mSv/yr (2 mrem/yr). Direct measurements have demonstrated that these calculated values are conservative.

Levels of tritium in the confinement building, at a nominal primary coolant concentration of 3.7×10^{10} Bq/L (1 Ci/L), are typically less than 0.01 DAC. The operating staff is in the building fewer than 1500 hr/yr, so this represents an individual dose commitment of less than 0.4 mSv/yr (40 mrem/yr). Bioassay data for the operating staff confirm that most exposures are well below 0.4 mSv/yr (40 mrem/yr). Other personnel are in the confinement building a much smaller fraction of the time, and their tritium exposures result in doses much less than 1 mrem/yr. Although reactor operators can be exposed to airborne sources of tritium during activities such as refueling, their doses would not normally exceed 1 mSv/yr (100 mrem/yr).

Airborne tritium levels can also be increased by abnormal or transient conditions. When the ventilation system for the NBSR was shut down for remediation over a 5-day period, the tritium levels slowly approached DAC values, and when an auxiliary cooling loop had excessive heavy water leakage, the local airborne tritium levels increased to 5 percent of the DAC.

The average radiation dose to facility workers from external exposure to radiation fields from 2001 to 2005 was less than 0.5 mSv/yr (50 mrem/yr), and the maximum annual exposures rarely exceed 5 mSv/yr (500 mrem/yr) during routine operations. Over that period, there were a total of 21 individual exposures that exceeded 5 mSv/yr (500 mrem/yr), and the maximum dose to an individual in any year ranged from 3.57 to 19.4 mSv (357 to 1940 mrem). Potential exposures to special populations, such as embryos or declared pregnant women, are very limited. Where such exposures could potentially exceed regulatory limits, added surveillance is provided and work is managed to further limit exposure to radiation and to radioactive materials.

Total annual exposure for the staff at NBSR over the last 5 years has ranged from 0.1 to 0.25 person-Sv (10 to 25 person-rem) for 676 to 914 monitored workers. Of those workers the number with measurable exposures (greater than 0.01 mSv [1 mrem]) ranged from 414 to 685. During a few earlier years that involved high-exposure maintenance and major upgrade activities, the yearly collective exposure to workers ranged from 0.18 to 0.22 person-Sv/yr (18 to 22 person-rem/yr).

2.2.8 Socioeconomic Factors

The staff reviewed the ER submitted by NIST and information obtained from county and city economic development staff. The following discussions describe the economy, population, and communities near the NBSR.

2.2.8.1 Housing

The NBSR is a national resource used by up to 2000 engineers and scientists for some part of their research every year. In 2002, the researchers came from 30 other Federal laboratories, 127 universities, 47 industrial laboratories, and 21 divisions and offices of NIST, and from all areas of the United States. According to a recent study by an interagency working group of the

Office of Science and Technology Policy (OSTP 2002), the NBSR is the highest performing and most used neutron facility in the United States.

Typically, visiting scientists and engineers will stay for 40 days, which corresponds to a reactor run cycle. Visiting scientists make their own housing arrangements while using the facility. No housing facility is provided by NIST for visiting scientists; however, there are over 50 hotels within 24 km (15 mi) of the site and many more in neighboring cities.

Although NIST is certainly a major employer in the Gaithersburg area, the local real estate market appears to be primarily driven by economic activity in the District of Columbia metropolitan center. The corridor connecting Washington, D.C., Baltimore, and Northern Virginia is home to over 8 million people, and the population within this area grew over 7 percent between 2000 and 2005. The existence of the NBSR within the NIST campus would appear to have little impact on local housing prices and rental rates.

Table 2-7 provides the number of housing units and housing unit vacancies for Gaithersburg and neighboring metropolitan areas within Montgomery County for 1990 and 2000. Gaithersburg, where the NBSR is located, had approximately 20,674 housing units in 2000, with a vacancy rate around 5 percent. Germantown, located to the northwest of Gaithersburg, had 21,568 housing units and a vacancy rate of 2 percent. Rockville and Montgomery Village, with a combined housing unit stock of just over 30,000 units, each has a vacancy rate of 3 percent.

In 1997, the Maryland legislature adopted legislation, commonly known as Smart Growth, aimed at slowing sprawl development in Maryland. The Smart Growth law targets State spending on roads, sewers, schools, and other public infrastructure in designated growth areas or priority funding areas. These areas include the land within the Baltimore and Washington beltways; established towns, cities, and rural villages; other existing and proposed communities above a minimum density; and industrial and employment areas. Although growth is not necessarily restricted in Montgomery County, the State of Maryland funnels significant funding into these designated growth areas, while no State funding is provided to development occurring outside of the designated growth areas. Given the land use and zoning designations in Montgomery County, there is currently a potential for another 241,000 housing units, of which 84 percent is in areas with existing or planned sewerage service (Maryland Department of Housing and Community Affairs 2001).

Table 2-8 contains data on population, estimated population, and annual population growth rates for Montgomery County. The population of Montgomery County has grown significantly in recent years and this level of growth is expected to continue throughout the next decade. This growth pattern is similar to other suburban counties surrounding the District of Columbia and also similar to overall growth rates for the State of Maryland.

Reactor and the Environment

Table 2-7. Total Occupied and Vacant (Available) Housing Units by County, 1990 and 2000

	1990	2000	Approximate Percentage Change
Gaithersburg			
Housing Units	16,059	20,674	29
Occupied Units	15,202	19,621	9
Vacant Units	857	1053	23
Germantown			
Housing Units	17,121	21,568	26
Occupied Units	15,784	20,893	32
Vacant Units	1337	375	-71
Rockville			
Housing Units	16,238	17,786	10
Occupied Units	15,660	17,247	21
Vacant Units	578	539	-7
Montgomery Village			
Housing Units	13,120	14,548	11
Occupied Units	12,284	14,142	15
Vacant Units	836	406	-51

Sources: U.S. Census Bureau (USCB 1990, 2000)

Table 2-8. Population Growth in Montgomery County, Maryland – 1980 to 2020

	Montgomery County		State of Maryland	
Year	Population	Percent Change (each decade)	Population	Percent Change (each decade)
1980	579,053	--	5,296,486	--
1990	757,027	10.8	4,781,468	10.9
2000	873,341	15.4	5,296,486	11.0
2010	975,000 (estimated)	11.6	5,904,025 (estimated)	11.5
2020	1,050,000 (estimated)	7.7	6,337,075 (estimated)	7.3

– No data available.
Sources: NIST 2004a; MDP 2007

2.2.8.2 Public Services

Public services, which include water supply, education, and transportation, are discussed below.

Water Supply

The WSSC, a co-operative utility, provides potable water to the City of Gaithersburg. The water is drawn from the Potomac River (the intake is upstream from Great Falls) and the Patuxent River. This water system operates with excess capacity with no expectations of problems in meeting future water demands of Gaithersburg. The average daily water demand on the system is 632 million L/day (167 million gallons per day [MGD]) with a peak demand of 1 billion L/day (267 MGD). The average demand is less than half the treatment capacity of 1.3 billion L/day (355 MGD) (WSSC 2005).

Transportation

As shown on Figure 2-3, I-270 forms the northeast boundary of the NIST campus and is a major commuter artery for workers in the Washington, D.C., metropolitan area living in Montgomery and Frederick Counties and other northern points. It is also a major truck route serving the area. Three arterial and collector roads abut the NIST campus (Figure 2-2). West Diamond Avenue, Quince Orchard Road, and Muddy Branch Road all serve the Gaithersburg area surrounding the NIST campus, providing truck routes serving the local economy (NIST 2004a).

A CSX rail line (CSX Transportation Corporation) paralleling the northeast boundary of the NIST campus carries goods through the region. At its closest point to the reactor, it is approximately 2 km (1.25 mi) away. This line also serves the MARC commuter train service that is used by people in northern Montgomery and Frederick Counties and other points north traveling to Washington, D.C. Shady Grove, the northernmost station for the MetroRail system is located approximately 5 km (3.0 mi) away from the reactor (NIST 2004a).

The I-270 Technology Corridor is a major research and development center in the State of Maryland. While some manufacturing does occur within this corridor, there are no significant manufacturing plants near the reactor, including no chemical plants or refineries. Mining and quarrying operations are limited to those associated with constructing new office buildings. A natural-gas pipeline lies 3.2 km (2 mi) to the south of the reactor, and a liquid petroleum/gas pipeline is located 1.6 km (1 mi) to the north (NIST 2004a).

Andrews Air Force Base, the nearest military base, is approximately 52 km (32.5 mi) away. A retired Nike missile site with its abandoned silos is located just to the south of the NIST campus. The three commercial airports within the region are IAD in northern Virginia; DCA in Virginia just across the Potomac River from Washington, D.C.; and BWI near Baltimore, Maryland. No associated normal air routes, holding patterns, or approach patterns are known to exist above the NIST campus. Montgomery Airpark is approximately 7 km (4.5 mi) to the northeast of the reactor. Its runway is oriented 140°/320° relative to magnetic north; that is, it is nearly perpendicular to the line between the reactor and the airfield. While the airfield can handle an aircraft as large as the Gulfstream 4, the largest aircraft typically using the field is the Falcon 900. There are approximately 140,000 annual take-offs and landings at this field. The airport has no known normal approach patterns. The typical air traffic in the general area is local air traffic, news aircraft, and an occasional military helicopter traversing the area (NIST 2004a).

Search of the National Transportation Safety Board database (covering 1962 to January 2007) revealed eight fatal accidents and 18 nonfatal accidents in the Gaithersburg area. One involved a hot-air balloon, while the remainder involved either airplanes or helicopters within Montgomery County. The following is a breakdown of the reported accidents:

- Fatal
  - Three occurred at the Montgomery County Airpark.
  - One occurred 1 km (0.6 mi) to the east of the Montgomery County Airpark.
  - One occurred 3.2 km (2.0 mi) to the northeast of the Montgomery County Airpark.
  - One occurred 8.1 km (5 mi) north of Montgomery County Airpark.
  - One occurred at an unspecified location within Montgomery County.
  - One involved a hot-air balloon.

- Non-Fatal
  - Eighteen non-fatal accidents occurred at the Montgomery County Airpark. The airpark is 7 km (4.5 mi) to the northeast of the NBSR. It is unlikely that the small aircraft flying into and out of this airpark pose any accident-related problems to the safe operation of the reactor (NTSB 2007).

2.2.8.3 Offsite Land Use

The NIST campus and general area within the 8-km (5-mi) circle surrounding the NBSR have a gently rolling topography. There are a few buildings within the area over three floors high, with the closest being the NIST Administration Building, which is approximately 1.25 km (0.75 mi) to the north of the NBSR. Other tall structures include several buildings in the Rio complex at the interchange of I-270 and I-370; these buildings are approximately 2.4 km (1.5 mi) to the east of the reactor (NIST 2004a).

The NBSR is located within the I-270 Technology Corridor, which is sited in the center of Montgomery County and constitutes the county's primary focus of economic and transportation activity. By 2015, 62 percent of the county's job growth and 51 percent of its household growth is expected to be within this area. The NBSR is surrounded by commercial buildings and suburban housing developments (Montgomery County 2005).

Most of Montgomery County is made up of urban and suburban/residential areas. In 2002, however, there were approximately 30,352 ha (75,000 ac) of land area devoted to agricultural use (out of the 128,285 ha [317,000 ac] total county land area). Just over 2995 ha (7400 ac) of Montgomery County are covered with water (USDA 2003).

2.2.8.4 Visual Aesthetics and Noise

The NBSR is in a suburban metropolitan area that is fairly densely populated. Most of the immediate area surrounding the NBSR lies on the campus of NIST. This area has laboratories and office buildings but no residential buildings. The closest permanent residences are more than 400 m (0.25 mi) directly to the east and directly to the west of the reactor (NIST 2004a).

There are several parks and recreation sites in close proximity to the NIST campus. Seneca Creek State Park is located to the northwest of the site and includes 2550 ha (6300 ac) along 22.5 km (14 mi) of Seneca Creek. East of the site are the Summit Hall Farm Park, Maple Lake Park, and Kelly Park. Just a couple miles further east in Derwood along Rock Creek is the much larger Agricultural History Farm Park, a 166-ha (410-ac) complex that connects with Rock Creek Regional Park. To the south of the site is the Muddy Branch Park, which includes an existing stream valley and network of trails beginning in Gaithersburg and connecting to the Potomac River (MCPPC 2006).

2.2.8.5 Demography

Demographic factors considered in this review included resident population, workforce, transient populations who stay temporarily to use NIST facilities, and the tax implications of the demographics.

Resident Population

The city of Gaithersburg surrounds the NIST campus (Figure 2-2). All of the area within the 2-km (1.25-mi) circle about the reactor and most of that within the 4-km (2.5-mi) circle are located in Gaithersburg. All of the town of Washington Grove and much of the city of Rockville also lie within the 8-km (5-mi) circle. Other unincorporated areas situated within the 8-km (5-mi) circle include Germantown, Montgomery Village, Darnestown, and North Potomac. According to 2000 Census data, the Germantown area was the seventh most populous place in Maryland with 55,419 residents; Gaithersburg was the tenth most populous with 52,613 residents, Rockville the fourteenth with 47,388 residents, and Montgomery Village was twenty-first at 38,051 residents. In terms of percentage growth of their populations between 1990 and 2000, this represents an increase of 34.7, 33.1, 5.7, and 17.8 percent, respectively. Table 2-5 presents the 1990 and 2000 Census data for these places.

Montgomery County is the most populous county in the State of Maryland. Much of this growth has occurred in the southern half of the county. Table 2-3 gives the 2000 to 2025 Census population and percentage change figures for the county. The populations within the 1-, 2-, 4-, 6-, and 8-km (0.6-, 1.2-, 2.5-, 3.7-, and 4.9-mi) radii about the reactor were estimated from the 2000 Census Population Counts by jurisdiction for the voting districts located within these encircled areas. For districts that are sited in more than one of the zones about the reactor, the percentage area located within each ring was estimated, and the population distribution within any one district was assumed linear with area. Table 2-1 gives the population estimates for each of the circles about the reactor for the years 2000, 2010, and 2025. These estimates are based on the voting district data.

Workforce

The service sector is the largest category of employment in Montgomery County, exceeding Federal, State and local government employment combined. This sector includes the following commercial areas: business and repair, personal services, entertainment and recreation, professional health services, professional education services, and other miscellaneous services. Over one-third of all jobs in Montgomery County are in the service industries. The second largest sector is retail trade, accounting for nearly one in five jobs in the county. The Federal government is the third largest employment sector, and the largest single employer, in the county. The locations of Federal installations in the county are provided in Figure 2-5 (Montgomery County 2005).

Major employers in lower Montgomery County include Marriott International, Lockheed Martin (the nation's largest defense contractor), the National Naval Medical Center, and Discovery Communications, which is building a new headquarters in downtown Silver Spring.

An economic recession in the early 1990s resulted in the loss of 20,000 jobs in Montgomery County. In 1992, an economic recovery began with employment growth continuing through 2003. Employment projections for through 2010 are included in Table 2-9.

In general, activities at NIST, including employment, contribute only a small share to the overall dynamics of the local economy.

Transient Populations

The NBSR is considered a national user facility, which means that scientists and engineers come from a number of different research institutions throughout the United States to use the facility on a temporary basis to complete their research. On average, approximately 1500 visiting scientists and engineers use the facility each year. Typically, visiting scientists and engineers will stay for 40 days, which corresponds to a reactor-run cycle. These visiting scientists are typically housed in local hotels (U.S. NRC 2007).

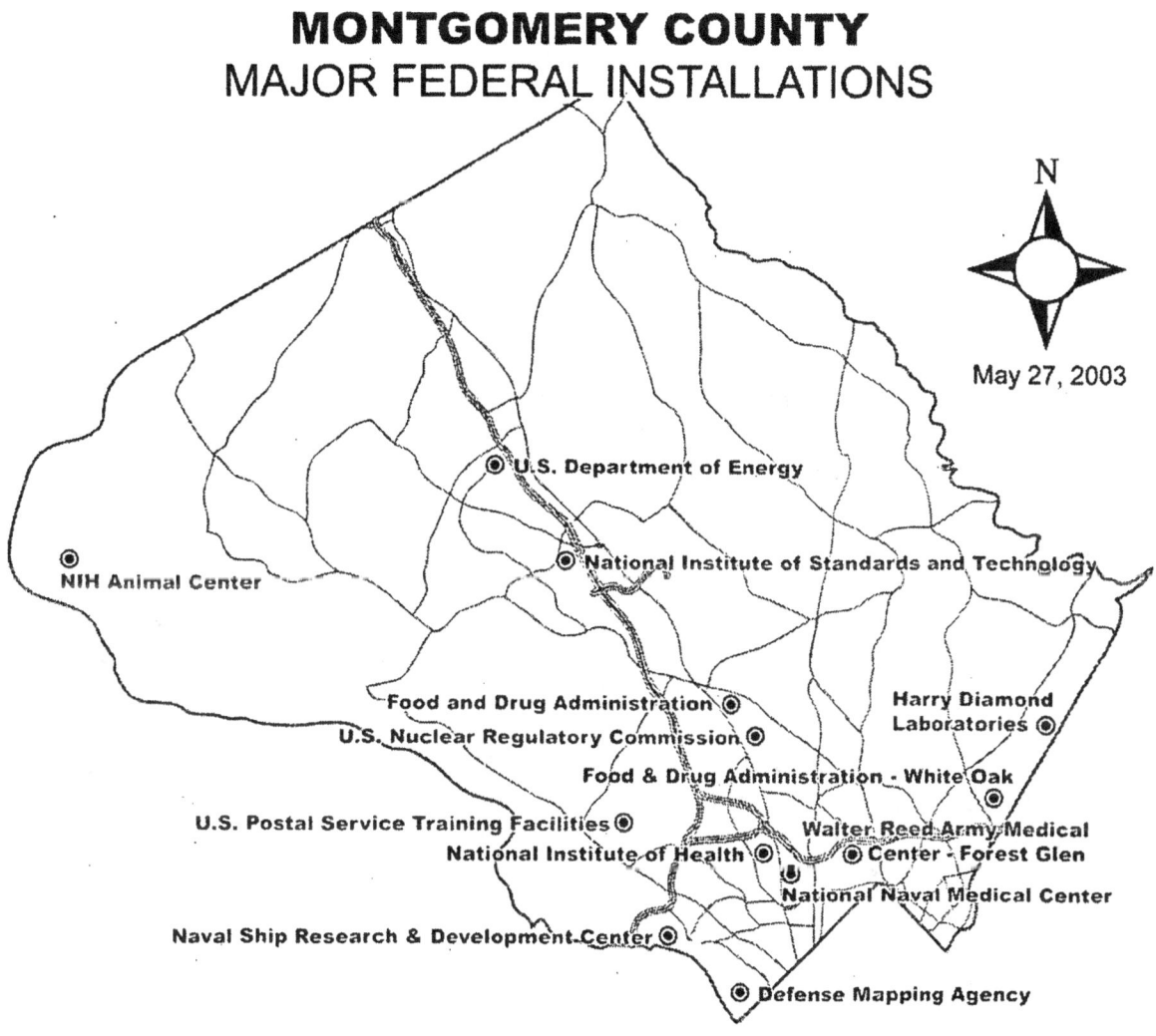

Figure 2-5. Major Federal Installations in Montgomery County, Maryland

Table 2-9. Number of Jobs in Montgomery County and the State of Maryland (2004 to 2010)

County/State	2004	2006	2008	2010	Average Annual Percent Change 2004-2010 (projected)	Percent Change, 1990-2001
Montgomery	505,000	530,000	549,000	565,000	0.4%	(-1.1%)
Maryland	2,764,110	2,876,013	--	--	--	14%

Sources: Montgomery County Parks and Planning Commission (2006); U.S. Bureau of Labor Statistics (2004, 2006)

Taxes

The NIST is a non-regulatory Federal agency of the U.S. Commerce Department within the Technology Administration. It is a tax-exempt research entity; therefore, there are no tax implications associated with the operation of the NIST Center for Neutron Research.

2.2.9 Historic and Cultural Resources

Although Maryland is rich in prehistoric and historic resources, according to the Maryland Historical Trust, the Maryland Inventory of Historic Properties does not have any record of known archeological sites or other historical properties within or immediately adjacent to the entire NIST campus (MDP 2006). There are no historic cemeteries surrounding the site. There are no Federally recognized tribes in Maryland, and the State of Maryland does not provide any official designation for tribal members. There are, however, several communities of indigenous people throughout the State who maintain an identity, including the Piscataway, the Nause-Waiwash, the Lenape, and the Lumbee. The closest historic district to the site is in Germantown (Jefferson Patterson Park and Museum 2006).

2.2.10 Related Federal Project Activities

The staff reviewed the possibility that activities of other Federal agencies might impact the issuance of a renewed operating license for the NBSR to NIST. Any such activities could result in cumulative environmental impacts and the possible need for a Federal agency to become a cooperating agency for preparation of this environmental impact statement (10 CFR 51.10(b)(2)).

Given the proximity of the NBSR to the District of Columbia, there are many Federal activities within the region (80 km [50 mi] radius of the NBSR). After considering the Federal activities in the vicinity of the NBSR, the staff determined there were no Federal project activities that would make it desirable for another Federal agency to become a cooperating agency for preparation of this environmental impact statement.

2.3 References

10 CFR Part 20. Code of Federal Regulations, Title 10, *Energy*, Part 20, "Standards for Protection Against Radiation."

10 CFR Part 51. Code of Federal Regulations, Title 10, *Energy*, Part 51, "Environmental Protection Regulations for Domestic Licensing and Related Regulatory Functions."

10 CFR Part 61. Code of Federal Regulations, Title 10, *Energy*, Part 61, "Licensing Requirements for Land Disposal of Radioactive Waste."

10 CFR Part 71. Code of Federal Regulations, Title 10, *Energy*, Part 71, "Packaging and Transportation of Radioactive Material."

10 CFR Part 100. Code of Federal Regulations, Title 10, *Energy*, Part 100, "Reactor Site Criteria."

40 CFR Part 58. Code of Federal Regulations, Title 40, *Protection of the Environment*, Part 58, "Ambient Air Quality Surveillance."

40 CFR Part 81. Code of Federal Regulations, Title 40, *Protection of the Environment*, Part 81, "Designations of Areas for Air Quality Planning Purposes."

40 CFR Part 190. Code of Federal Regulations, Title 40, *Protection of the Environment*, Part 190, "Environmental Radiation Protection Standards for Nuclear Power Operations."

Endangered Species Act. 16 USC 1531 et seq.

Jefferson Patterson Park and Museum. 2006. Accessed on the Internet at http://www.jefpat.org/diagnostic/Prehistoric_Ceramic_Web_Page/Prehistoric_Prehistory.htm on January 4, 2006.

Maryland Department of Housing and Community Affairs. 2001. "Technical Supplement to the Montgomery County, Maryland Housing Policy: Montgomery County - The Place to Call Home." Rockville, Maryland. Accessed on the Internet at http://www.co.mo.md.us/hca.

Maryland Department of Natural Resources (MDNR). 2002. "Maryland's Coastal Program: What is the Coastal Zone?" Accessed on the Internet at http://www.dnr.state.md.us/bay/czm/coastal_zone.html on December 1, 2006.

Reactor and the Environment

Maryland Department of Natural Resources (MDNR). 2004. "Current and Historical Rare, Threatened, and Endangered Species of Montgomery County, Maryland." Accessed on the Internet at http://www.dnr.state.md.us/wildlife/index.asp on January 16, 2006.

Maryland Department of Planning (MDP). 2006. Letter, dated February 17, 2006, from Elizabeth J. Cole, Administrator, Project Review and Compliance. ML070380446.

Maryland Department of Planning, Maryland State Data Center (MDP). 2007. "Population Projections by Type for All Counties Historic Census 1970 to 2000, projected 2005 to 2030." Accessed on the Internet at http://www.mdp.state.md.us/msdc/dw_Popproj.htm on October 16, 2007.

Montgomery County Department of Environmental Protection. 2006. "The Muddy Branch Watershed." Accessed on the Internet at http://www.montgomerycountymd.gov/deptmpl.asp?url=/content/dep/csps/watersheds/csps/html/muddy.asp on February 6, 2006.

Montgomery County Government website and information. 2005. Accessed on the Internet at http://www.montgomerycountymd.gov in December 2005.

Montgomery County Parks and Planning Commission (MCPPC). 2006. Accessed on the Internet at http://www.mc-mncppc.org/research/index.shtm in January 2006.

National Climatic Data Center (NCDC). 2005a. *2004 Local Climatological Data Annual Summary With Comparative Data - Washington, D.C., Ronald Regan National Airport (DCA)*. Asheville, North Carolina.

National Climatic Data Center (NCDC). 2005b. *2004 Local Climatological Data Annual Summary With Comparative Data - Washington, D.C., Dulles International Airport (IAD)*. Asheville, North Carolina.

National Environmental Policy Act of 1969 (NEPA). 42 USC 4321, et seq.

National Institute of Standards and Technology (NIST). 2002. *National Institute of Standards and Technology Reactor. Operations Report #54. January 1, 2001 to December 31, 2001*. NIST, Gaithersburg, Maryland.

National Institute of Standards and Technology (NIST). 2003. *National Institute of Standards and Technology Reactor. Operations Report #55. January 1, 2002 to December 31, 2002*. NIST, Gaithersburg, Maryland.

National Institute of Standards and Technology (NIST). 2004a. *Environmental Report for License Renewal for the National Institute of Standards and Technology Reactor-NBSR.* NBSR-16, NISTIR 7105, NIST, Gaithersburg, Maryland.

National Institute of Standards and Technology (NIST). 2004b. *National Institute of Standards and Technology Reactor. Operations Report #56. January 1, 2003 to December 31, 2003.* NIST, Gaithersburg, Maryland.

National Institute of Standards and Technology (NIST). 2004c. *Technical Specifications for License Renewal for the National Institute of Standards and Technology Reactor-NBSR.* NBSR-15, NISTIR 7104, NIST, Gaithersburg, Maryland.

National Institute of Standards and Technology (NIST). 2005. *National Institute of Standards and Technology Reactor. Operations Report #57. January 1, 2004 to December 31, 2004.* NIST, Gaithersburg, Maryland.

National Institute of Standards and Technology (NIST). 2006. *National Institute of Standards and Technology Reactor. Operations Report #58. January 1, 2005 to December 31, 2005.* NIST, Gaithersburg, Maryland.

National Transportation and Safety Board (NTSB). 2007. Accessed on the Internet at http://www.ntsb.gov/ntsb/query.asp on February 1, 2007.

Newman, M.E. 2005. "Deer Immunocontraception at NIST." Accessed on the Internet at http://www.nist.gov/public_affairs/factsheet/deer.htm on November 3, 2007.

Office of Science and Technology Policy (OSTP). 2002. *Report on the Status and Needs of Major Neutron Scattering Facilities and Instruments in the United States.* Washington, D.C.

Ramsdell, J.V. 2005. *Tornado Climatology of the Contiguous United States.* NUREG/CR-4461, Rev. 1, U.S. Nuclear Regulatory Commission, Washington, D.C.

U.S. Bureau of Labor Statistics (USBLS). 2004. "Bureau of Labor Statistics Data." Accessed on the Internet at http://data.bls.gov/labjava/outside.jsp?survey=1a on July 22, 2004.

U.S. Bureau of Labor Statistics (USBLS). 2006. "Site Source." Accessed on the Internet at http://www.dllr.state.md.us/lmi/ on October 2006.

U.S. Census Bureau (USCB). 1990. "U.S. Bureau of Census, 1990 Census of Population and Housing, Table DP-1." Accessed on the Internet at http://factfinder.census.gov on April 19, 2004.

Reactor and the Environment

U.S. Census Bureau (USCB). 2000. "U.S. Bureau of Census, Census 2000, State and County Quick Facts." Accessed on the Internet at http://quickfacts.census.gov/qfd/ on April 19, 2004.

U.S. Department of Agriculture (USDA). 2003. "NASS Fact Finders for Agriculture." Accessed on the Internet at http://www.nass.usda.gov/md/Montgomery2003Profile.pdf in January 2006.

U.S. Environmental Protection Agency (U.S.EPA). 1989. *User's Guide for the COMPLY Code.* EPA 520/1-89-003, EPA Office of Radiation and Indoor Air, Washington, D.C.

U.S. Environmental Protection Agency (U.S.EPA). 2005. "Air Quality Index Report." Accessed on the Internet at http://www.epa.gov/air/data/repsst.html?st~MD~Maryland on January 20, 2006.

U.S. Nuclear Regulatory Commission (U.S.NRC). 1982. *Final Environmental Statement Related to License Renewal and Power Increase for the National Bureau of Standards Reactor.* NUREG-0877. Washington, D.C.

U.S. Nuclear Regulatory Commission (U.S.NRC). 2007. *Summary of Site Audit to Support the License Renewal Review for the National Bureau of Standards Reactor (NBSR) at the National Institute of Standards and Technology.* ML070370061.

Washington Suburban Sanitary Commission (WSSC). 2005. Accessed on the Internet at http://www.wsscwater.com/index.cfm during December 2005.

3.0 Environmental Impacts of Operation

This chapter addresses the environmental impacts related to operation during the license renewal term of the National Bureau of Standards Reactor (NBSR) located on the National Institute of Standards and Technology (NIST) site in Montgomery County, Maryland.

There are substantial differences between the NBSR and commercial power reactors; however, the types of environmental issues addressed in this chapter are similar, and in many cases, the environmental impacts from continued operation of the NBSR can be informed by analyses discussed in the *Generic Environmental Impact Statement for License Renewal of Nuclear Plants* (GEIS), NUREG-1437, Volumes 1 and 2 (U.S. NRC 1996, 1999).[a] Therefore, where appropriate, the GEIS analyses are used as a basis for evaluating the environmental impacts of continued operation of the NBSR. The environmental impacts of operating the NBSR during the license renewal term are presented in the following sections.

Unless otherwise indicated, information in the following sections was adapted from the Environmental Report (ER) submitted by NIST for renewal of the NBSR operating license (OL) (NIST 2004), and was independently verified by the staff. Additional information was obtained by the staff during the site audit (U.S. NRC 2007); appropriate citations will be made for other sources. Section 3.1 addresses issues applicable to the NBSR cooling system. Section 3.2 addresses the radiological impacts of normal operation, and Section 3.3 addresses issues related to the socioeconomic impacts of normal operation during the license renewal term. Section 3.4 addresses issues related to historic and archaeological resources, while Section 3.5 discusses the impacts of license renewal-term operations on terrestrial and aquatic resources, including threatened and endangered species. Section 3.6 discusses cumulative impacts, and Section 3.7 summarizes the results of the evaluation of environmental issues related to operation during the license renewal term. References are listed in Section 3.8.

3.1 Cooling System

The NBSR primary cooling is provided by a closed system containing heavy water (D_2O). The primary system is connected to a secondary cooling system containing light water (H_2O) via a plate-type heat exchanger. The secondary system consists of a plume abatement cooling tower that uses make-up water from a municipal utility as needed and discharges blowdown to the sanitary sewer system. The potential for leakage between the primary and secondary systems is carefully monitored, and if contaminants were transferred to the secondary system, they would be removed and managed with other radiological liquid waste, as necessary.

(a)   The GEIS was originally issued in 1996. Addendum 1 to the GEIS was issued in 1999. Hereafter, all references to the "GEIS" include the GEIS and its Addendum 1.

Operation

The NBSR cooling system does not discharge water to an open body of water, and no impacts on surface water quality or on biota that would normally inhabit rivers or lakes would be expected.

For all of the following environmental issues associated with cooling tower systems, the staff concluded the impacts from operation of nuclear power reactors are SMALL, and additional mitigation measures are not likely to be sufficiently beneficial to be warranted as described in the GEIS (U.S. NRC 1996):

- Discharge of sanitary wastes and minor chemical spills
- Discharge of chlorine or other biocides
- Discharge of other metals in wastewater
- Cooling tower impacts on crops and ornamental vegetation
- Cooling tower impacts on native plants
- Bird collisions with cooling towers
- Microbiological organisms (occupational health)
- Noise.

Other environmental issues associated with cooling system operation evaluated in GEIS Section 4.3 are not considered to be applicable to the NBSR. By comparison to power reactors, the NBSR operates at a substantially lower power level (from a factor of 75 to 150 or more). The cooling tower technology employed at NBSR is similar in principle to those at power reactor facilities, but the scale is likewise reduced.

In its ER, the NIST did not identify any information for cooling system operations indicating potential for impacts greater than, or different in nature from, those discussed in the GEIS (NIST 2004). The staff reviewed applicable information related to cooling system impacts during its independent review of the ER, the staff's site visit, the scoping process, and its evaluation of other available information. Based on the analysis and findings of the GEIS for similar cooling system technology and the fact the NBSR operates at a substantially lower power level than commercial power reactors, the staff concludes the impacts are bounded by the impacts for commercial power reactors (i.e., SMALL), and no additional mitigation is warranted.

3.2 Radiological Impacts

Radiological issues related to impacts at nuclear power plants applicable to the NBSR include radiation exposures to the public and occupational radiation exposures. For these issues, the staff concluded in the GEIS Section 4.6 (U.S. NRC 1996) that the impacts from commercial power reactor operations are SMALL, and additional plant-specific mitigation measures are not likely to be sufficiently beneficial to be warranted. The staff reviewed applicable information related to radiological impacts on workers and members of the public during its independent review of the ER, the scoping process, the staff's site visit, and its evaluation of other available information. Because the NBSR operates at a substantially lower power level than a commercial power reactor, there is no expectation that radiological impacts of operating the NBSR would differ from, or exceed, those discussed in the GEIS in Section 4.6 (U.S. NRC 1996). Radiation exposures to the public as a result of operating the NBSR during the license renewal term are expected to continue at current levels associated with normal operations, as discussed in Section 2.2.7 of this EIS. This includes exposures to radionuclides in airborne and liquid effluents as well as direct radiation. Likewise, projected maximum occupational doses during the license renewal term are within the range of doses experienced during normal operations or normal maintenance outages, and would continue to be well below regulatory limits.

Radiological impacts from ongoing research projects at the NIST Center for Neutron Research are the only unique activities associated with normal operation of the NBSR. Doses to members of the public from research activities are included in the radiological impacts of reactor operation based on monitoring of effluents and various environmental media, and they represent a small fraction of the dose from reactor operations (NIST 2004). Radiological doses to research staff working in the laboratories are monitored and are typically lower than those to reactor operations staff, as discussed in Section 2.2.7 of this EIS. Therefore, they would be well below regulatory standards and within the bounds of the GEIS Section 4.6 analysis (U.S. NRC 1996). Consequently, the staff concludes the radiological impacts associated with operation during a renewal term would be SMALL, and no additional mitigation measures beyond the existing control program are warranted.

3.3 Socioeconomic Impacts

Socioeconomic impacts considered include those on housing availability, public services (utilities and transportation), land use, and environmental justice.

3.3.1 Housing Impacts

SMALL impacts result when no discernible change in housing availability occurs, changes in rental rates and housing values are similar to those occurring statewide, and no housing construction or conversion is required to meet new demand. Impacts are considered MODERATE when there is discernible but short-lived reduction in available housing units because of project-induced migration. Impacts are considered LARGE when project-related housing demands result in very limited housing availability and would increase rental rates and housing values well above normal inflation.

Appendix C of the GEIS (U.S. NRC 1996) presents a population characterization method based on two factors, sparseness and proximity. Sparseness measures population density within 32 km (20 mi) of the site, and proximity measures population density and city size within 80 km

Operation

(50 mi). Each factor has categories of density and size (U.S. NRC 1996, Table C.1), and a matrix is used to rank the population category as low, medium, or high (U.S. NRC 1996, Figure C.1).

In 2000, the population living in Montgomery County, where the NBSR is located, was estimated to be approximately 873,341. This total converts to a population density of about 680 persons/km^2 (1775 persons/mi^2). This population density falls into the GEIS sparseness Category 4 (i.e., having greater than or equal to 46 persons/km^2 [120 persons/mi^2]). In addition, the District of Columbia metropolitan area has a population of approximately 4.8 million and is located about 32 km (20 mi) southeast of the site (NIST 2004). Applying the GEIS proximity measures (U.S. NRC 1996), NBSR is classified as being located in a high-population area.

The NRC has concluded the impacts on housing availability are expected to be of SMALL significance at commercial power reactors located in a high-population area where growth-control measures are not in effect. The NBSR site is located in a high-population area. In 1997, the Maryland legislature adopted legislation, commonly known as Smart Growth, aimed at slowing sprawl development in Maryland. The Smart Growth law targets State spending on roads, sewers, schools, and other public infrastructure in designated growth areas or priority funding areas. Growth is not necessarily restricted in Montgomery County; however, the State directs significant resources into designated growth areas, while no State funding is provided for development occurring outside of the designated growth areas. Given the land-use and zoning designations in Montgomery County, there is currently a potential for another 241,000 housing units, of which 84 percent is expected to be in areas with existing or planned sewerage service (Maryland Department of Housing and Community Affairs 2001); therefore, the growth control measures in place would not appear to significantly restrict future housing growth around the site.

The demand for housing units in the Montgomery County region could be met with the construction of new housing. As a result, the NRC staff concludes the impacts on housing would be SMALL, and mitigation measures would be neither necessary nor effective. Based on this review, including interviews with local real estate agents, the staff concludes the impact on housing during the license renewal term would be SMALL, and no mitigation is warranted.

3.3.2 Public Services: Public Utilities

Impacts on public utility services are considered SMALL if there is little or no change in the ability of the system to respond to the level of demand, and thus there is no need to add capital facilities. Impacts are considered MODERATE if overtaxing of service capabilities occurs during periods of peak demand. Impacts are considered LARGE if existing levels of service (e.g., water or sewer services) are substantially degraded and additional capacity is needed to meet ongoing demands for services. The staff believes the only potential significant impacts on public utilities are impacts on public water supplies.

Analysis of impacts on the public water supply system considered both plant demand and plant-related population growth. Section 2.2.8.2 describes the NBSR-permitted withdrawal rate and actual use of water. NIST does not plan to undertake any major change in activities during the license renewal term at the NBSR, and none of the activities would require staffing that would exceed the NBSR's current level of staffing, so plant demand would not change beyond current demands (NIST 2004). Thus, the staff concludes the impact of increased water use resulting from the potential increase in employment is SMALL, and no mitigation is warranted.

3.3.3 Public Services: Transportation

As described in Section 2.2.8 of this EIS, the road network around the NIST campus is well established and in heavy use by commuters within Montgomery County who travel to and from the District of Columbia and other surrounding large cities. Operations during the license renewal term of the NBSR would be expected to have SMALL impacts on transportation, and no mitigation is warranted.

3.3.4 Offsite Land Use

Consistent with the definitions from Section 4.7.4 of the GEIS to define the magnitude of land-use changes as a result of plant operation during the license renewal term, the following terms are used to analyze land-use impacts:

SMALL – Little new development and minimal changes to an area's land-use pattern

MODERATE – Considerable new development and some changes to the land-use pattern

LARGE – Large-scale new development and major changes in the land-use pattern.

There would be no expected population growth as a result of renewing the OL for the NBSR facility. Consequently, the staff concludes that population changes resulting from license renewal are likely to result in SMALL offsite land-use impacts, and no mitigation is warranted.

3.3.5 Environmental Justice

Environmental justice refers to a Federal policy requiring Federal agencies to identify and address, as appropriate, disproportionately high and adverse human health or environmental effects of its actions on minority[a] or low-income populations. The memorandum accompanying

---

(a)   The NRC Guidance for performing environmental justice reviews defines "minority" as American Indian or Alaskan Native, Asian or Pacific Islander, Black not of Hispanic Origin, or Hispanic (U.S. NRC 2004).

Operation

Executive Order 12898 (59 FR 7629) directs Federal executive agencies to consider environmental justice under the National Environmental Policy Act of 1969 (NEPA). The Council on Environmental Quality (CEQ) has provided guidance for addressing environmental justice (CEQ 1997). Although the Executive Order is not mandatory for independent agencies, the Commission has voluntarily committed to undertake environmental justice reviews; the Commission has finalized its approach for considering environmental justice reviews in its Policy Statement (69 FR 52040). Specific guidance is provided in NRC Office of Nuclear Reactor Regulation Office Instruction LIC-203, Revision 1, "Procedural Guidance for Preparing Environmental Assessments and Considering Environmental Issues," issued in May 2004 (U.S. NRC 2004).

The staff examined the geographic distribution of minority and low-income populations within Montgomery County and neighboring counties, employing the 2000 Census data (USCB 2000) for low-income populations and minority populations. For the purpose of the staff's review, a minority population is defined to exist if the percentage of each minority, or aggregated minority category within the census tract or block group[a] potentially affected by the license renewal of NBSR exceeds the corresponding percentage of minorities in the entire State of Maryland by 20 percent or if the corresponding percentage of minorities within the census tract or block group is at least 50 percent. A low-income population is defined to exist if the percentage of low-income population within a census tract or block group exceeds the corresponding percentage of low-income population in the entire State of Maryland by 20 percent, or if the corresponding percentage of low-income population within a census tract or block group is at least 50 percent. The minority population in the State of Maryland makes up 35 percent of the population, and the low-income population makes up 8.8 percent of the total population in the State.

Applying the LIC-203 (U.S. NRC 2004) criterion of "more than 20 percent greater," the census block groups containing low-income populations appeared to be primarily in the urban centers around the District of Columbia and Baltimore, Maryland, with only two block groups identified in Montgomery County and two more identified in Frederick County to the north.

Minority population block groups are present in Montgomery County and all adjacent counties; however, the concentrations of these minority populations are found in the urban centers within and surrounding Baltimore and the District of Columbia.

(a)   A census block group is a combination of census blocks, which are statistical subdivisions of a census tract. A census block is the smallest geographic entity for which the U.S. Census Bureau (USCB) collects and tabulates decennial census information. A census tract is a small, relatively permanent statistical subdivision of counties delineated by local committees of census data users in accordance with USCB guidelines for the purpose of collecting and presenting decennial census data. Census block groups are subsets of census tracts (USCB 2001).

With the locations of minority and low-income populations identified, the staff proceeded to evaluate whether any of the environmental impacts of the proposed action could affect these populations in a disproportionately high and adverse manner. Based on staff guidance (U.S. NRC 2004), air, land, and water resources within and around the NBSR site were examined. The pathways through which the environmental impacts associated with NBSR license renewal can affect human populations are discussed in each associated section. The staff found no unusual resource dependencies or practices such as subsistence agriculture, hunting, or fishing through which minority and/or low-income populations could be disproportionately highly and adversely affected. In addition, the staff did not identify any location-dependent, disproportionately high and adverse impacts affecting these minority and low-income populations. The staff concludes offsite impacts from NBSR to minority and low-income populations would be SMALL, and no mitigation is warranted.

3.4 Historic and Archaeological Resources

Section 106 of the National Historic Preservation Act requires Federal agencies to take into account the effects of their undertakings on historic properties. The Section 106 historic preservation review process is covered in regulations issued by the Advisory Council on Historic Preservation at 36 CFR Part 800. As a starting point, renewal of the NBSR OL could potentially affect historic properties that may be located at the site. However, the Maryland Inventory of Historic Properties does not have any records of known archaeological sites or other historic properties within or immediately adjacent to the NBSR or the entire NIST campus (MDP 2006). NRC staff consulted the Maryland Historic Trust regarding the potential renewal of the OL for the NBSR because the staff ultimately determined, in accordance with 36 CFR 800.3(a)(1), that renewal would be an activity that does not have the potential to cause effects on historic properties. Operation of the NBSR, as planned under the application for license renewal, would protect undiscovered historic or archaeological resources on the NIST site because the undeveloped natural landscape and vegetation would remain undisturbed, and access to the site would remain restricted. Therefore, the staff concludes the environmental impacts on cultural resources associated with operation during a renewal term would be SMALL, and no additional mitigation measures are warranted. As a Federal agency, activity conducted by NIST that could result in disturbing land on the NIST campus would conform with the requirements of the National Historic Preservation Act.

3.5 Ecology

Ecological impacts considered include those for aquatic and terrestrial resources, as well as threatened and endangered species.

3.5.1 Aquatic Ecology

The closed-cycle secondary cooling system has its intake via municipal-water supply lines, and blowdown is discharged to the sanitary sewer system. Surface water and groundwater are not used as process water, and process water is not discharged to the surface or groundwater. Therefore, no impacts on aquatic biota as a result of impingement, entrainment, heat, or chlorination are expected to occur.

Overall impacts to the aquatic biota are expected to be SMALL, and no mitigation is warranted.

3.5.2 Terrestrial Ecology

The NBSR and associated facilities are located in an industrial complex on the NIST campus. Because of the highly industrialized nature of the facility, it is not expected that terrestrial biota will be impacted from continued operation. Fogging and icing as a result of cooling tower drift and evaporation are not expected other than in the immediate vicinity of the cooling towers. Bird collisions are not expected to occur on either the cooling towers or at the buildings housing these facilities. There have been no visible impacts to vegetation from cooling tower drift recorded in the last 20 years (U.S. NRC 2007). The average annual precipitation of 104 cm (41 in.) is distributed more or less evenly throughout the year, and it is expected that it will wash the deposited drift from vegetative surfaces and prevent accumulation of high salt levels in the soil (NOAA 2006). Impact on surrounding terrestrial vegetation from the cooling tower drift is expected to be small.

Overall impacts to the terrestrial biota are expected to be SMALL, and no mitigation is warranted.

3.5.3 Threatened and Endangered Species

Section 7(a)(2) of the Endangered Species Act states that Federal agencies are to consult with the U.S. Fish and Wildlife Service (FWS) to ensure any agency action is not likely to jeopardize the continued existence of any endangered species or threatened species or result in the destruction or adverse modification of habitat of such species. Although no threatened or endangered species are known to occur on the NIST campus, official consultation has been initiated with the FWS.

Aquatic

There are no known threatened and endangered aquatic species in the vicinity of the NIST campus. No impacts to threatened and endangered aquatic species are expected; therefore, the impacts on aquatic threatened and endangered species are expected to be SMALL, and no mitigation is warranted.

Terrestrial

There is suitable habitat for the small whorled pogonia (*Isotria medeoloides*) on the NIST campus. The NBSR and associated facilities are located within an industrial complex on the campus. There are no planned construction activities as part of license renewal requiring any additional habitat removal (U.S. NRC 2007). Because of the highly industrialized nature of the facility and the fact no construction is planned, it is not expected the small whorled pogonia would be impacted from continued operation. Overall, impacts to terrestrial threatened and endangered species is expected to be SMALL, and no additional mitigation is warranted.

3.6 Cumulative Impacts of Operations During the License Renewal Term

The cumulative effects of impacts were considered for operation of the cooling system, radiological doses, socioeconomics, historic and archaeological resources, and ecology.

3.6.1 Cumulative Impacts Resulting from Operation of the Plant Cooling System

The geographic area affected by operation of the NBSR cooling system is confined largely to the NIST site. The plume abatement cooling tower minimizes the potential for substantial offsite impacts; therefore, the opportunity for cumulative impacts on nearby facilities is small. Effluents released to the municipal sanitary sewer system from cooling tower operations represent a small fraction of the site's total volume, and they are monitored to maintain concentrations of radiological or hazardous materials well within regulatory limits for discharges to public treatment facilities. NRC and EPA regulatory standards were established at levels that account for contributions from multiple sources to releases of radiological or hazardous materials, thereby minimizing the potential for cumulative adverse impacts to public facilities that process the effluents. Therefore, the staff concludes the cumulative impacts resulting from continued operation of the NBSR cooling system would be SMALL, and no additional mitigation is warranted.

Operation

3.6.2 Cumulative Radiological Impacts

The EPA and NRC established radiological dose limits for protection of the public and workers from both near-term and cumulative impacts of exposure to radiation and radioactive materials. Those dose limits are codified in 40 CFR Part 190 and 10 CFR Part 20. For the purpose of this analysis, the area within an 80-km (50-mi) radius of the NIST site was considered. The NIST conducts a radiological environmental monitoring program (REMP) at the NBSR site to measure radiation and radioactive materials from all sources, including the reactor and associated research facilities (NIST 2006). Historically, these measurements have remained at relatively constant low levels and provide no indication of cumulative effects over time. Other laboratories within the NIST campus may also use radioactive materials. Radiological exposures to workers at NIST are monitored to ensure they do not exceed regulatory standards. In addition, the NRC staff concluded that impacts of radiation exposure to the public and workers (occupational) from operation of the NBSR during the license renewal term are small. The NRC and the State of Maryland would regulate any future actions in the vicinity of the NIST site that could contribute to cumulative radiological impacts; none are contemplated at this time.

Therefore, the staff concludes that cumulative radiological impacts of continued operations of the NBSR would be SMALL, and no additional mitigation is warranted.

3.6.3 Cumulative Socioeconomic Impacts

The analyses of socioeconomic impacts presented in Section 3.3 already incorporate cumulative impact analysis. For instance, the impact of the total number of additional housing units that may be needed can only be evaluated with respect to the total number that will be available in the affected area. Given that all license renewal socioeconomic impacts associated with the NBSR are SMALL, the staff concluded these impacts would not result in significant cumulative impacts on potentially affected socioeconomic resources, and no mitigation is warranted.

3.6.4 Cumulative Impacts on Historic and Archaeological Resources

The Maryland Inventory of Historic Properties does not have any record of known archaeological sites or other historic properties within or immediately adjacent to the NBSR or the entire NIST campus. Given that all license renewal historical and archaeological impacts associated with the NBSR are deemed to be SMALL, the staff concluded these impacts would not result in significant cumulative impacts on historic and archaeological resources, and no mitigation is warranted.

3.6.5 Cumulative Impacts on Ecology Including Threatened and Endangered Species

There are no known threatened and endangered aquatic species in the vicinity of the NIST campus. There is suitable habitat for one Federally listed terrestrial specie on the NIST campus; however, the NBSR and associated facilities are located within an industrial complex on the NIST campus, and no new construction is planned as part of license renewal within the industrial complex or elsewhere on the NIST campus. Therefore, the staff determined continued operation at the plant site will not have a detectable contribution to the cumulative, regional impacts on threatened or endangered aquatic and terrestrial species. The effects are SMALL, and no mitigation is warranted.

3.7 Summary of Impacts of Operations During the License Renewal Term

The NBSR is a small, non-power test reactor located in a wing of a building within the industrial complex on the NIST campus. It uses municipal water for make-up water, and blowdown is discharged directly to the sanitary sewer system. The number of employees is small in relation to the population of the surrounding community. Radiological impacts are minimized by meeting applicable regulations for releases, monitoring, and doses to workers and the public, including implementing an ALARA program. Therefore, the staff concludes the potential environmental impact of renewal-term operations of the NBSR would be SMALL, and no additional mitigation is warranted.

3.8 References

10 CFR Part 20. Code of Federal Regulations, Title 10, *Energy*, Part 20, "Standards for Protection Against Radiation."

10 CFR Part 51. Code of Federal Regulations, Title 10, *Energy*, Part 51, "Environmental Protection Regulations for Domestic Licensing and Related Regulatory Functions."

36 CFR Part 800. Code of Federal Regulations, Title 36, *Parks, Forests, and Public Property*, Part 000, "Protection of Historic Properties."

40 CFR Part 190. Code of Federal Regulations, Title 40, *Protection of Environment*, Part 190, "Environmental Protection Standards for Nuclear Power Operations."

59 FR 7629. Executive Order 12898, "Federal Actions to Address Environmental Justice in Minority and Low-Income Populations." *Federal Register*. Vol. 59, No. 32. February 16, 1994.

Operation

69 FR 52040. "Policy Statement on the Treatment of Environmental Justice Matters in the NRC Regulatory and Licensing Actions." *Federal Register*. Vol. 69, No. 163. August 24, 2004.

Council on Environmental Quality (CEQ). 1997. *Environmental Justice: Guidance Under the National Environmental Policy Act*. Executive Office of the President, Washington, D.C.

Endangered Species Act. 16 USC 1531 et seq.

Maryland Department of Housing and Community Affairs. 2006. "Technical Supplement to the Montgomery County, Maryland Housing Policy: Montgomery County – The Place to Call Home," Report available upon request at www.co.mo.md.us/hca. July 2001. Rockville, Maryland.

Maryland Department of Planning (MDP). 2006. Letter, dated February 17, 2006, from Elizabeth J. Cole, Administrator, Project Review and Compliance to Katie A. Cort, PNNL. ML070380446.

National Historic Preservation Act of 1966 (NHPA). 16 USC 470, et seq.

National Institute of Standards and Technology (NIST). 2004. *Environmental Report for License Renewal for the National Institute of Standards and Technology Reactor-NBSR*. NBSR-16, NISTIR 7105, NIST, Gaithersburg, Maryland.

National Institute of Standards and Technology (NIST). 2006. *National Institute of Standards and Technology Reactor. Operations Report #58. January 1, 2005 to December 31, 2005*. NIST, Gaithersburg, Maryland.

National Oceanic and Atmospheric Administration. 2006. "Climate at a Glance." Accessed on the Internet at http://climvis.ncdc.noaa.gov/cgi-bin/cag3/hr-display3.pl on February 6, 2006.

U.S. Census Bureau (USCB). 2000. "Census 2000 Summary File I (SF-1) 100 Percent Data." Accessed on the Internet at http://factfinder.census.gov/servlet/DatasetMainPageServlet?_ds_name=DEC_2000_SF3_U&_program=DEC&_lang=en in January 2003.

U.S. Census Bureau (USCB). 2001. "Glossary – Definition and Explanations – decennial census terms." Accessed on the Internet at http://landview.census.gov/dmd/www/advglossary.html in January 2003.

U.S. Nuclear Regulatory Commission (U.S.NRC). 1996. *Generic Environmental Impact Statement for License Renewal of Nuclear Plants*. NUREG-1437, Volumes 1 and 2, Washington, D.C.

U.S. Nuclear Regulatory Commission (U.S.NRC). 1999. *Generic Environmental Impact Statement for License Renewal of Nuclear Plants, Main Report,* "Section 6.3 – Transportation, Table 9.1, Summary of findings on NEPA issues for license renewal of nuclear power plants, Final Report." NUREG-1437, Volume 1, Addendum 1, Washington, D.C.

U.S. Nuclear Regulatory Commission (U.S.NRC). 2004. "Procedural Guidance for Preparing Environmental Assessments and Considering Environmental Issues." LIC-203, Revision 1, May 24, 2004. ML033550003.

U.S. Nuclear Regulatory Commission (U.S.NRC). 2007. *Summary of Site Audit to Support the License Renewal Review for the National Bureau of Standards Reactor (NBSR) at the National Institute of Standards and Technology.* ML070370061.

4.0 Environmental Impacts of Postulated Accidents

The potential impacts of accidents at the National Bureau of Standards Reactor (NBSR) located at the National Institute of Standards and Technology (NIST) during the license renewal term are presented in the following sections. Environmental issues associated with postulated accidents at nuclear power reactors are discussed in the *Generic Environmental Impact Statement for License Renewal of Nuclear Plants (GEIS)*, NUREG-1437, Volumes 1 and 2 (U.S. NRC 1996, 1999).[a] There are substantial differences between the NBSR and commercial power reactors; however, the types of environmental issues addressed in this chapter are similar and, in many cases, the environmental impacts from continued operation of the NBSR can be informed by analysis discussed in the GEIS. Therefore, where appropriate, the GEIS analyses are used as a basis for evaluating the environmental impacts of continued operation of the NBSR. The GEIS includes a determination of whether the analysis of a particular environmental issue could be applied to all commercial power reactors and whether additional mitigation measures would be warranted.

This chapter describes the environmental impacts from postulated accidents at the NBSR considered for the license renewal term. Section 4.1 presents postulated accidents, Section 4.2 addresses severe accident mitigation alternatives, and references are listed in Section 4.3.

Unless otherwise indicated, information in the following sections was adapted from the Environmental Report (ER) (NIST 2004a) and the Final Safety Analysis Report (SAR) (NIST 2004b) submitted by NIST for renewal of the NBSR operating license (OL), and was independently verified by the NRC staff. Additional information was obtained by the staff during the site audit (U.S. NRC 2007); appropriate citations are made for other sources.

4.1 Postulated Facility Accidents

Two classes of accidents are evaluated for commercial power plants in the GEIS. These are referred to as design-basis accidents (DBAs) and severe accidents. Corresponding accidents evaluated for the NBSR are discussed in the following sections.

(a)   The GEIS was originally issued in 1996. Addendum 1 to the GEIS was issued in 1999. Hereafter, all references to the "GEIS" include the GEIS and Addendum 1.

4.1.1 Design-Basis Accidents

To receive U.S. Nuclear Regulatory Commission (NRC) approval to operate a nuclear reactor, an applicant must submit a SAR as part of the application. The SAR presents the design criteria and design information for the proposed reactor and comprehensive data on the proposed site. The SAR also discusses hypothetical accident scenarios as well as the safety features present in the facility to prevent and mitigate accidents. The NRC staff reviews the application to determine whether the facility design meets the Commission's regulations and requirements. The facility design includes, in part, the reactor design and its anticipated response to an accident.

Design-basis accidents are postulated and evaluated to ensure the reactor can withstand normal and abnormal transient conditions and a broad spectrum of postulated accidents, without undue hazard to the health and safety of the public. A number of the postulated accidents are not expected to occur during the life of the facility but are evaluated to establish the design basis for the preventive and mitigative safety systems of the facility. The acceptance criteria for DBAs are described in Title 10 of the Code of Federal Regulations (CFR) Part 50 and 10 CFR Part 100.

The environmental impacts of DBAs were evaluated during the initial licensing process for the NBSR, and the ability of the facility to withstand these accidents was demonstrated to be acceptable before NRC issued the OL. The results of these evaluations are found in license documentation such as the staff's Safety Evaluation Report (SER), the licensee's updated Final SAR, and this environmental impact statement (EIS). The licensee is required to maintain the acceptable design and performance criteria throughout the life of the facility, including the license renewal period. The consequences of accidents are evaluated for the hypothetical maximally exposed individual, and as such, changes in the facility environment would not affect these evaluations. Renewal of the OL requires accident consequences remain acceptable and aging management programs are in effect. Therefore, the environmental impacts as calculated for DBAs over the life of the facility, including the license renewal period, should not differ significantly from initial licensing assessments. Accordingly, the design of the facility relative to DBAs during the license renewal period is considered to remain acceptable, and the environmental impacts of those accidents were not examined further in the GEIS.

The NBSR includes many inherent, passive safety features, some of which would preclude the types of reactor accidents commonly evaluated for nuclear power plants. The prompt neutron lifetime is relatively long as a result of heavy water moderation, and the reactivity coefficients of void and temperature are negative. The reactor operates in a low-temperature, un-pressurized condition and does not have a large stored energy content. The cooling system is designed to retain coolant in the event of a loss of water from the reactor vessel and to supply emergency coolant flow to the fuel elements and the reactor core without operator intervention. Design-basis accidents evaluated in the NBSR Final SAR (NIST 2004b) included start-up, maximum reactivity insertion, loss of flow, fuel handling, and loss of coolant. The evaluations demonstrate that none of these accidents would result in a safety hazard to the public or to the environment.

The NIST ER (NIST 2004a) did not identify any information relevant to accident impacts associated with the renewal of the NBSR OL. In addition, the staff has not identified any concerns during the staff's independent review of the ER, the scoping process, the staff's site visit, and its evaluation of other available information. With respect to nuclear power reactors, the Commission determined the environmental impacts of DBAs are of SMALL significance for all plants because the plants are designed to successfully withstand these accidents. The power levels of commercial power reactors are of the order of 100 times greater than that of the NBSR and are expected to bound the environmental impacts of the DBAs for the NBSR. Therefore, the staff concludes there are no impacts of DBAs during the license renewal term that exceed or differ substantially from those discussed in the GEIS and further mitigation is not warranted.

4.1.2 Severe Accidents

Severe nuclear accidents include events that could result in damage to the reactor core, whether or not there are serious offsite consequences, and they are considered separately from DBAs. The GEIS assessed the impacts of severe accidents at commercial power reactors during the license renewal period, using the results of existing analyses and site-specific information to conservatively predict the environmental impacts of severe accidents for each plant during the license renewal period.

The only severe accident identified for the NBSR is discussed in the facility Final SAR (NIST 2004b). That event, designated the maximum hypothetical accident (MHA), is one in which all coolant flow through a single fuel element is blocked while the reactor is operating at full power. Such an event is highly unlikely because the NBSR is a closed system with upward flow. However, if the flow in an element is blocked during full-power operation, it is possible some melting of the cladding could occur with a resultant release of fission products into the primary coolant. In evaluating the consequences of the MHA, it was conservatively assumed the entire blocked element's cladding would melt and release fission products into the primary cooling system. Analysis of consequences following the MHA in the NBSR Final SAR estimated the total whole-body gamma dose to a person standing at the site boundary 24 hours a day for 30 days would be 7 mrem and the iodine dose to the thyroid would be negligible. Those consequences would be well below limits specified for DBAs in 10 CFR Part 100.

The staff reviewed information concerning severe accidents during its independent review of the ER, the scoping process, the site visit, and its evaluation of other available information and concludes that the impacts of severe accidents at commercial power reactors as discussed in the GEIS would bound any potential accidents at the NBSR.

As part of the GEIS analysis, the probability weighted consequences of atmospheric releases, fallout onto open bodies of water, releases to groundwater, and societal and economic impacts from severe accidents were determined to be SMALL and further mitigation is not warranted. The power levels of commercial power reactors are of the order of 100 times greater than that of the NBSR and are expected to bound the environmental impacts of severe accidents for the NBSR. No design changes are proposed for the NBSR, and no severe accident mitigation analysis is required.

4.2 References

10 CFR Part 50. Code of Federal Regulations, Title 10, *Energy*, Part 50, "Domestic Licensing of Production and Utilization Facilities."

10 CFR Part 51. Code of Federal Regulations, Title 10, *Energy*, Part 51, "Environmental Protection Regulations for Domestic Licensing and Related Regulatory Functions."

10 CFR Part 100. Code of Federal Regulations, Title 10, *Energy*, Part 100, "Reactor Site Criteria."

National Institute of Standards and Technology (NIST). 2004a. *Environmental Report for License Renewal for the National Institute of Standards and Technology Reactor-NBSR*. NBSR-16, NISTIR 7105, NIST, Gaithersburg, Maryland.

National Institute of Standards and Technology (NIST). 2004b. *Safety Analysis Report (SAR) for License Renewal for the National Institute of Standards and Technology Reactor-NBSR*. NBSR-14, NISTIR 7102, NIST, Gaithersburg, Maryland.

U.S. Nuclear Regulatory Commission (U.S.NRC). 1996. *Generic Environmental Impact Statement for License Renewal of Nuclear Plants*. NUREG-1437, Vols. 1 and 2, Washington, D.C.

U.S. Nuclear Regulatory Commission (U.S.NRC). 1999. *Generic Environmental Impact Statement for License Renewal of Nuclear Plants, Main Report*, "Section 6.3 – Transportation, Table 9.1, Summary of findings on NEPA issues for license renewal of nuclear power plants, Final Report." NUREG-1437, Volume 1, Addendum 1, Washington, D.C.

5.0 Environmental Impacts of the Uranium Fuel Cycle and Solid Waste Management

This chapter addresses the environmental impacts related to the nuclear fuel cycle and solid waste management related to operating the National Bureau of Standards Reactor (NBSR) at the National Institute of Standards and Technology (NIST) site during the license renewal term. In many cases, the impacts resulting from renewal of the NBSR operating license can be extrapolated from previous analyses for commercial power reactors in the *Generic Environmental Impact Statement for License Renewal of Nuclear Plants (GEIS)*, NUREG-1437, Volumes 1 and 2 (U.S. NRC 1996, 1999).[a] For power reactors, environmental impacts from the supporting uranium fuel cycle were evaluated based on a model 1000-MWe (megawatts of electric power) light-water cooled reactor (LWR) operating at an annual capacity factor of 80 percent. Accounting for the efficiency of producing electric power from thermal power and the capacity factor of 80 percent, the power level of a commercial reactor is on the order of 100 times greater than the NBSR. The results of the analyses are listed in 10 CFR 51.51(b), Table S–3, "Table of Uranium Fuel Cycle Environmental Data," and in 10 CFR 51.52(c), Table S–4, "Environmental Impact of Transportation of Fuel and Waste to and from One Light-Water-Cooled Nuclear Power Reactor." The staff's analysis of the radiological impact from radon-222 and technetium-99 releases are addressed separately from those listed in the tables (U.S. NRC 1996, 1999). The principal radon releases occur during uranium mining and milling operations and as emissions from mill tailings, whereas the principal technetium-99 releases occur from gaseous diffusion uranium enrichment facilities.

The NBSR differs from a commercial power reactor in several respects: (1) it uses highly enriched uranium (HEU) fuel compared to the low enriched uranium (LEU) fuel used in commercial power reactors, (2) the reactor core is cooled and moderated by heavy water (D_2O) rather than light water (H_2O), and (3) it operates at a much lower power level (20 MWt [megawatts of thermal power] compared to about 3000 MWt for a typical power reactor). Therefore, the impacts from the uranium fuel cycle for the NBSR could differ from those for a commercial power reactor, particularly those resulting from use of HEU fuel rather than LEU fuel. However, the staff's conclusion for most types of environmental impacts would not be altered if the analysis were to be based on the operation of the NBSR after applying appropriate scaling factors for the power output (of the order of 100 times smaller) compared to a model LWR.

There are substantial differences between the NBSR and commercial power reactors; however, the types of environmental issues addressed in this chapter are similar, and in many cases, the

---

(a)   The GEIS was originally issued in 1996. Addendum 1 to the GEIS was issued in 1999. Hereafter, all references to the "GEIS" include the GEIS and its Addendum 1.

environmental impacts from continued operation of the NBSR can be informed by analyses discussed in the GEIS. Environmental impacts from the NBSR uranium fuel cycle are discussed in the following section.

5.1 The Uranium Fuel Cycle

The Environmental Report (ER) submitted by NIST (NIST 2004) did not specifically address environmental impacts from the uranium fuel cycle related to the renewal of the NBSR operating license. The staff performed an independent review of the ER, the scoping process, and conducted a site visit.

A brief description of the staff's review, the conclusions, and a discussion of their applicability to the NBSR for each of the issues follows:

- Offsite radiological impacts (individual effects from other than the disposal of spent fuel and high level waste)

  Offsite impacts of the uranium fuel cycle have been considered by the Commission in Table S–3 of 10 CFR 51.51(b). Accounting for the differences between the model LWR and the NBSR, including the differences in power level and fuel enrichment, radiological environmental impacts on individuals from radioactive gaseous and liquid releases (including radon-222 and technetium-99) are expected to be small.

- Offsite radiological impacts (collective effects)

  Offsite impacts (collective effects) of the uranium fuel cycle have been considered by the Commission in Table S–3 of 10 CFR 51.51(b). Accounting for the differences between the model LWR and the NBSR, including the differences in fuel power level and enrichment, collective radiological environmental impacts on populations from radioactive gaseous and liquid releases (including radon-222 and technetium-99) are expected to be small.

- Offsite radiological impacts (spent fuel and high-level waste [HLW] disposal)

  The fuel used for the NBSR is owned by the U.S. Department of Energy (DOE), and DOE is responsible for its storage, processing, and disposal. The radiological impacts from management of spent fuel and high level waste, including interim storage and disposal of highly enriched test reactor fuel, have also been evaluated separately (U.S.DOE 1995, 2002).

Despite the current uncertainty with respect to licensing of a HLW repository, some judgment as to the implications of offsite radiological impacts of spent fuel and high-level waste disposal should be made. The staff concludes these impacts would be sufficiently small that the option of extending the NBSR operating license should be preserved. Based on the volume of spent fuel generated during the license renewal period at the NBSR and its total radionuclide content, the impacts from disposal of NBSR spent fuel relative to those from a commercial power reactor are considered to be small.

At this time, there are no facilities for permanent disposal of HLW. The Nuclear Waste Policy Act of 1982 defined the goals and structure of a program for permanent, deep geologic repositories for HLW and unreprocessed spent fuel. Under this Act, the DOE is responsible for developing permanent disposal capacity for the spent fuel and other high-level nuclear wastes. At the present time, DOE, as directed by the Congress, is investigating a site in Yucca Mountain, Nevada, for a possible disposal facility. An HLW repository would be built and operated by DOE and licensed by the NRC. The Commission believes (10 CFR 51.23(a)) there is reasonable assurance that at least one mined geological repository will be available in the first quarter of the 21st Century and that, within 30 years beyond the licensed life of operation for any reactor, sufficient repository capacity will be available to dispose of the reactor's HLW and spent fuel generated up to that time.

The Commission has independently, in a separate proceeding (i.e., the Waste Confidence Proceeding), made a finding that there is:

> ...reasonable assurance that, if necessary, spent fuel generated in any reactor can be stored safely and without significant environmental impacts for at least 30 years beyond the licensed life for operation (which may include the term of a revised license) of that reactor at its spent fuel storage basin, or at either onsite or offsite independent spent fuel storage installations (54 FR 39767).

The Commission has committed to review this finding at least every 10 years. In its most recent review, the Commission concluded that experience and developments since 1990 were not such that a comprehensive review of the Waste Confidence Decision was necessary at this time (64 FR 68005). Accordingly, the Commission reaffirmed its findings of insignificant environmental impacts cited above. This finding is codified in the Commission's regulations at 10 CFR 51.23(a). The staff relies on the Waste Confidence Rule, but for completeness has elected to include in this EIS information related to the storage and maintenance of fuel in a spent fuel pool.

As stated earlier, the spent fuel from the NBSR is stored at the NIST and then shipped to the DOE Savannah River Site for reprocessing or shipment to a permanent repository. By comparison to power reactors, the NBSR operates at substantially lower power levels (by a factor of 75 to 150 or more) and the quantity of fuel for the NBSR reactor is likely be to smaller by the same factor. Therefore, the staff concludes the relatively small quantities involved in the extended period of operation do not bring into question the Commission's Waste Confidence Decision.

- Nonradiological impacts of the uranium fuel cycle

Based on the relative quantities of fuel and total fissile material required for the NBSR relative to those for a model LWR, the nonradiological impacts of the uranium fuel cycle resulting from the renewal of an operating license for NBSR are considered to be small.

- Low-level waste storage and disposal

The comprehensive regulatory controls in place and the low public doses being achieved at reactors ensure the radiological impacts to the environment will remain small during the term of a renewed license. Because low-level waste is transported regularly for treatment as necessary and disposal, the maximum additional onsite land required for low-level waste storage during the term of a renewed license and associated impacts will be small. Nonradiological impacts on air and water will be negligible. The radiological and nonradiological environmental impacts of long-term disposal of low-level waste from any reactor are small. In addition, the staff concludes that there is reasonable assurance sufficient low-level waste disposal capacity will be made available when needed for facilities to be decommissioned consistent with requirements promulgated by the U.S. Nuclear Regulatory Commission (NRC).

- Mixed waste storage and disposal

The comprehensive regulatory controls and the facilities and procedures in place ensure proper handling and storage, as well as negligible doses and exposure to toxic materials for the public and the environment for all reactors. License renewal will not increase the small risk to human health and the environment posed by mixed waste at all reactors. The radiological and nonradiological environmental impacts of long-term disposal of mixed waste from any reactor are small. In addition, the staff concludes there is reasonable assurance sufficient mixed waste disposal capacity will be made available when needed for facilities to be decommissioned consistent with NRC decommissioning requirements.

- Onsite spent fuel

The onsite radiological impacts from interim storage of NBSR spent fuel are considered small. The fuel used for the NBSR is owned by the DOE, and DOE is responsible for its storage, processing, and disposal. Because the NBSR regularly ships spent fuel offsite for storage, the onsite impacts of managing it are expected to remain small.

- Nonradiological waste

No changes to nonradiological waste generation are anticipated for the NBSR during the license renewal period. Facilities and procedures are in place to ensure continued proper handling and disposal, and the impacts from managing the wastes are considered to be small.

- Transportation

The impacts of transporting spent fuel from the model LWR to a single repository, such as Yucca Mountain, Nevada were found to be consistent with the impact values contained in 10 CFR 51.52(c), Summary Table S–4. The fuel used for the NBSR is owned by the DOE, and DOE is responsible for its processing and disposal. Spent fuel from the NBSR is transported from the NIST site to the DOE Savannah River Site for storage. The radiological impacts from management of spent fuel and HLW, including transportation of highly enriched test reactor fuel, have also been evaluated separately (U.S. DOE 1995, 2002). Based on the volume of spent fuel generated during the license renewal period at the NBSR and its total radionuclide content, the impacts from transporting NBSR spent fuel relative to those from a commercial power reactor are considered to be small.

Based on the foregoing, the staff concludes there are no significant environmental impacts related to the uranium fuel cycle; therefore, the impacts are SMALL, and no mitigation is warranted.

5.2 References

10 CFR Part 51. Code of Federal Regulations, Title 10, *Energy*, Part 51, "Environmental Protection Regulations for Domestic Licensing and Related Regulatory Functions."

54 FR 39767. "Proposed Waste Confidence Decision." *Federal Register*, Vol. 54, pp. 39,767. September 28, 1989.

64 FR 68005. "Waste Confidence Decision Review: Status." *Federal Register*, Vol. 64, No. 233, pp. 68,005-68,007. December 6, 1999.

Uranium Fuel Cycle

National Institute of Standards and Technology (NIST). 2004. *Environmental Report for License Renewal for the National Institute of Standards and Technology Reactor-NBSR.* NBSR-16, NISTIR 7105, NIST, Gaithersburg, Maryland.

U.S. Department of Energy (U.S.DOE). 1995. *Department of Energy Programmatic Spent Nuclear Fuel Management and Idaho National Engineering Laboratory Environmental Restoration and Waste Management Programs Final Environmental Impact Statement.* DOE/EIS-0203-F, U.S. Department of Energy, Office of Environmental Management, Idaho Operations Office, Idaho Falls, Idaho.

U.S. Department of Energy (U.S.DOE). 2002. *Final Environmental Impact Statement for a Geologic Repository for the Disposal of Spent Nuclear Fuel and High-Level Radioactive Waste at Yucca Mountain, Nye County, Nevada.* DOE/EIS-0250F, U.S. Department of Energy, Washington, D.C.

U.S. Nuclear Regulatory Commission (U.S.NRC). 1996. *Generic Environmental Impact Statement for License Renewal of Nuclear Plants.* NUREG-1437, Vols. 1 and 2, Washington, D.C.

U.S. Nuclear Regulatory Commission (U.S.NRC). 1999. *Generic Environmental Impact Statement for License Renewal of Nuclear Plants, Main Report,* "Section 6.3 – Transportation, Table 9.1, Summary of findings on NEPA issues for license renewal of nuclear power plants, Final Report." NUREG-1437, Volume 1, Addendum 1, Washington, D.C.

6.0 Environmental Impacts of Decommissioning

Environmental impacts from decommissioning research and test reactors are addressed in the *Final Generic Environmental Impact Statement on Decommissioning of Nuclear Facilities*, NUREG-0586, published in August 1988 (U.S. NRC 1988). A supplement to NUREG-0586 was published to update information regarding commercial power reactors (U.S. NRC 2002). Although information in the original NUREG would be most directly applicable to estimating decommissioning impacts for the National Bureau of Standards Reactor (NBSR), updated information in Supplement 1 to NUREG-0586 regarding waste management, transportation, or other areas is useful for this analysis.

The incremental environmental impacts associated with decommissioning activities resulting from continued operation of commercial power reactors during the license renewal term were evaluated in the *Generic Environmental Impact Statement for License Renewal of Nuclear Plants (GEIS)*, NUREG-1437, Volumes 1 and 2 (U.S. NRC 1996, 1999).[a]

There are substantial differences between the NBSR and commercial power reactors; however, the types of environmental issues addressed in this chapter are similar, and in many cases the environmental impacts from continued operation of the NBSR can be informed by analyses discussions in the GEIS. Therefore, where appropriate, the GEIS analyses are used as a basis for evaluating the environmental impacts of continued operation of the NBSR. The environmental impacts related to decommissioning from operating the NBSR during the license renewal term are presented in the following sections.

6.1 Decommissioning

Decommissioning issues related to the NBSR following the renewal term are discussed in the following sections. The Environmental Report (ER) submitted by the National Institute of Standards and Technology (NIST) did not identify information associated with impacts of decommissioning of the NBSR (NIST 2004). In addition, the staff has not identified any additional relevant information concerning impacts during its independent review of the ER, the scoping process, the staff's site visit, or its evaluation of other available information. Therefore, the staff concludes there are no impacts related to these issues beyond those discussed in either the GEIS for license renewal (U.S. NRC 1996, 1999) or NUREG-0586 and Supplement 1 (U.S. NRC 1988, 2002).

(a) The GEIS was originally issued in 1996. Addendum 1 to the GEIS was issued in 1999. Hereafter, all references to the "GEIS" include the GEIS and its Addendum 1.

Decommissioning

A brief description of the staff review, the GEIS conclusions, and a discussion of their applicability to the NBSR for each of the issues follows:

- Radiation doses. Based on information in the GEIS, the Commission found that doses to the public will be well below applicable regulatory standards regardless of which decommissioning method is used. Occupational doses would increase no more than 1 person-rem (0.01 person-Sv) caused by buildup of long-lived radionuclides during the license renewal term.

- Waste management. Based on information in the GEIS, the Commission found that decommissioning at the end of a 20-year license renewal period would generate no more solid wastes than at the end of the current license term. No increase in the quantities of Class C or greater than Class C wastes would be expected.

- Air quality. Based on information in the GEIS, the Commission found that air quality impacts of decommissioning are expected to be negligible either at the end of the current operating term or at the end of the license renewal term.

- Water quality. Based on information in the GEIS, the Commission found that the potential for significant water quality impacts from erosion or spills is no greater whether decommissioning occurs after a 20-year license renewal period or after the original 40-year operation period, and measures are readily available to avoid such impacts.

- Ecological resources. Based on information in the GEIS, the Commission found that decommissioning after either the initial operating period or after a 20-year license renewal period is not expected to have any direct ecological impacts.

- Socioeconomic Impacts. Based on information in the GEIS, the Commission found that decommissioning would have some short-term socioeconomic impacts. The impacts would not be increased by delaying decommissioning until the end of a 20-year relicense period, but they might be decreased by population and economic growth.

For all of these issues, the staff concluded that the impacts from decommissioning reactors are SMALL, and additional mitigation measures are not likely to be sufficiently beneficial to be warranted. Because the NBSR is expected to contain smaller quantities of radioactive and hazardous materials than commercial power reactors at the end of its license renewal term, the impacts from decommissioning the NBSR would be well within the range of those discussed for commercial power reactors (U.S. NRC 1996, 1999, 2002).

For the subjects discussed above, the staff has not identified any relevant information during its independent review of the ER, the staff's site visit, the scoping process, or its evaluation of other available information. Therefore, the staff concludes radiation dose, waste management, air quality, water quality, ecological resource, and socioeconomic impacts associated with decommissioning the NBSR following the license renewal term are bounded by those discussed in the GEIS.

6.2 References

10 CFR Part 51. Code of Federal Regulations, Title 10, *Energy,* Part 51, "Environmental Protection Regulations for Domestic Licensing and Related Regulatory Functions."

National Institute of Standards and Technology (NIST). 2004. *Environmental Report for License Renewal for the National Institute of Standards and Technology Reactor-NBSR.* NBSR-16, NISTIR 7105, NIST, Gaithersburg, Maryland.

U.S. Nuclear Regulatory Commission (U.S.NRC). 1988. *Final Generic Environmental Impact Statement on Decommissioning of Nuclear Facilities.* NUREG-0586, Washington, D.C.

U.S. Nuclear Regulatory Commission (U.S.NRC). 1996. *Generic Environmental Impact Statement for License Renewal of Nuclear Plants.* NUREG-1437, Vols. 1 and 2, Washington, D.C.

U.S. Nuclear Regulatory Commission (U.S.NRC). 1999. *Generic Environmental Impact Statement for License Renewal of Nuclear Plants, Main Report,* "Section 6.3 – Transportation, Table 9.1, Summary of findings on NEPA issues for license renewal of nuclear power plants, Final Report." NUREG-1437, Volume 1, Addendum 1, Washington, D.C.

U.S. Nuclear Regulatory Commission (U.S.NRC). 2002. *Generic Environmental Impact Statement on Decommissioning of Nuclear Facilities. Supplement 1 Regarding the Decommissioning of Nuclear Power Reactors. Final Report.* NUREG-0586, Supplement 1, Vols. 1 and 2. Office of Nuclear Reactor Regulation, Washington, D.C.

7.0 Environmental Impacts of the Alternatives

This chapter examines the potential environmental impacts associated with alternatives to the proposed action. The alternatives considered are (1) denying the renewal of the operating license (OL) (i.e., the no-action alternative) for the National Bureau of Standards Reactor (NBSR) at the National Institute of Standards and Technology (NIST), (2) constructing a new reactor and associated support facilities to replace the capabilities of the NBSR, and (3) performing work currently conducted at the NBSR at alternative existing research facilities. For the third alternative, the NRC staff determined that comparable alternative facilities do not exist in the United States. The NBSR is the nation's only cold-neutron source with the range of instrumentation that can meet the needs of the U.S. neutron-scattering science program, and it has the only very high inelastic cold-neutron spectrometer, spin echo, and backscattering instruments in the United States. In addition, it is very difficult for U.S. scientists to secure research time at potentially suitable foreign facilities, such as the Institut Laue-Langevin facility in France; the Paul Scherrer Institut laboratory in Switzerland; or the Forshungsreakter Munchen reactor in Germany. For these reasons, the staff did not consider these foreign research facilities to be viable alternatives to the NBSR. Consequently, the third alternative was not considered further.

Using the U.S. Nuclear Regulatory Commission's (NRC) established license renewal evaluation framework for commercial power reactors ensures a thorough evaluation of the impacts of renewal of the OL for the NBSR. The *Generic Environmental Impact Statement for License Renewal of Nuclear Plants (GEIS)*, NUREG-1437, Volumes 1 and 2 (U.S. NRC 1996, 1999)[a] was written specifically for use in the renewal of operating licenses for commercial power reactors. In conducting the staff review of the NIST application, the NRC was informed by certain GEIS features including the use of the three-level standard of significance. In following the precedent of the GEIS and site-specific supplemental license renewal environmental impact statements (EISs), environmental issues have been evaluated using a three-level standard of significance – SMALL, MODERATE, or LARGE – developed using the Council on Environmental Quality guidelines and set forth in the footnotes to Table B-1 of 10 CFR Part 51, Subpart A, Appendix B:

> SMALL – Environmental effects are not detectable or are so minor that they will neither destabilize nor noticeably alter any important attribute of the resource.

> MODERATE – Environmental effects are sufficient to alter noticeably, but not to destabilize important attributes of the resource.

(a)   The GEIS was originally issued in 1996. Addendum 1 to the GEIS was issued in 1999. Hereafter, all references to the "GEIS" include the GEIS and its Addendum 1.

LARGE – Environmental effects are clearly noticeable and are sufficient to destabilize important attributes of the resource.

7.1   No-Action Alternative

The NRC regulations implementing National Environmental Policy Act of 1969 (NEPA) specify the no-action alternative be discussed in an NRC EIS (10 CFR Part 51, Subpart A, Appendix A(4)). For license renewal, the no-action alternative refers to a scenario in which the NRC would not renew the OL for the NBSR. The NIST would then decommission the NBSR and the associated facilities covered under the OL at some future time.

The NIST would be required to comply with NRC decommissioning requirements whether or not the NBSR OL is renewed. If the OL is renewed, decommissioning activities could be postponed for up to an additional 20 years. If the OL is not renewed, decommissioning activities would be conducted by the NIST according to the requirements in 10 CFR 50.82(b).

The environmental impacts of the no-action alternative are summarized in Table 7-1 and are discussed in the following sections. Implementation of the no-action alternative would also have certain positive impacts in that adverse environmental impacts associated with current operation of the NBSR, however small they may be, would be eliminated.

7.1.1   Land Use

Temporary changes in onsite land use could occur during decommissioning, including addition or expansion of staging and laydown areas or construction of temporary buildings and parking areas. No offsite land-use changes are expected as a result of decommissioning. Following decommissioning, the land occupied by the NBSR would likely be retained by NIST for other purposes. The staff concludes the impacts of the no-action alternative on land use would be SMALL.

7.1.2   Aquatic and Terrestrial Resources

Land disturbance during decommissioning is expected to be minimal and would result in relatively short-term ecological impacts that could be mitigated using best management practices. The land is expected to recover naturally. No impacts to threatened or endangered species as a result of decommissioning activities are anticipated. Overall, the staff concludes the impacts to aquatic and terrestrial resources associated with the no-action alternative would be SMALL.

Table 7-1. Summary of Environmental Impacts of the No-Action Alternative

Impact Category	Impact	Comment
Land Use	SMALL	Onsite impacts expected to be temporary. No offsite impacts expected.
Aquatic and Terrestrial Resources	SMALL	Impacts are expected to be minimal, temporary, and largely mitigatable using best management practices.
Water Use and Quality	SMALL	Water use will decrease. Water quality unlikely to be adversely affected during decommissioning.
Air Quality	SMALL	Greatest impact is likely to be from fugitive dust; impact can be mitigated by good management practices.
Waste	SMALL	LLW will be disposed of at DOE or licensed facilities. A permanent disposal facility for HLW is not currently available.
Human Health	SMALL	Radiological doses to workers and members of the public are expected to be within regulatory limits. Occupational injuries are possible, but injury rates at nuclear reactors are below the U.S. average industrial rate.
Socioeconomics	SMALL	Slight decrease in employment.
Aesthetics	SMALL	Small positive impact from eventual removal of buildings and structures. Some noise impact during decommissioning operations.
Historic and Archaeological Resources	SMALL	Minimal impact on land utilized during reactor operations. Land occupied by the NBSR would likely be retained by NIST for other purposes.
Environmental Justice	SMALL	Minimal impact.

7.1.3 Water Use and Quality

Cessation of plant operations would result in a reduction in water use because reactor cooling would no longer be required. As plant staff size decreases, the demand for potable water would be expected to decrease as well. Overall, the staff concludes that the water use and quality impacts of decommissioning would be SMALL.

7.1.4 Air Quality

Decommissioning activities that can adversely affect air quality include dismantlement of systems and equipment, demolition of buildings and structures, and operation of internal combustion engines. The most likely adverse impact would be the generation of fugitive dust. Best management practices, such as seeding and wetting, can be used to minimize the generation of fugitive dust. Overall, the staff concludes the air quality impacts associated with decommissioning activities would be SMALL.

7.1.5 Waste

Decommissioning activities would result in the generation of radioactive and nonradioactive waste. Low-level radioactive waste (LLW) would be transferred to the U.S. Department of Energy (DOE) or disposed of in a facility licensed by the NRC or a state with authority delegated by the NRC. Recent advances in volume reduction and waste processing have significantly reduced waste volumes. A permanent repository for high-level radioactive waste (HLW) is not currently available. The NRC has made a generic determination that, if necessary, spent fuel generated in any reactor can be stored safely and without significant environmental impacts for at least 30 years beyond the licensed life for operation (which may include the term of a revised or renewed license) of that reactor at its spent fuel storage basin or at either onsite or offsite independent spent fuel storage installations (10 CFR 51.23(a)). Disposal of nonradioactive waste would be at offsite disposal facilities with appropriate permits. Overall, the staff concludes the waste impacts associated with the no-action alternative would be SMALL.

7.1.6 Human Health

Radiological doses to occupational workers during decommissioning and collective doses to members of the public and to the maximally exposed individual as a result of decommissioning activities would be well below the limits in 10 CFR Part 20. Occupational injuries to workers engaged in decommissioning activities are possible; however, historical injury and fatality rates at nuclear reactors have been lower than the average U.S. industrial rates. Overall, the staff concludes the human health impacts associated with the no-action alternative would be SMALL.

7.1.7 Socioeconomics

If the NBSR ceases operation, there would be a decrease in employment. However, impacts would be minimal because NBSR employment levels are relatively small and numerous other employers are in the Washington, D.C., metropolitan area. The no-action alternative would result in the loss of NBSR payrolls approximately 20 years earlier than if the OL were renewed. Overall, the staff concludes the socioeconomic impacts resulting from implementation of the no-action alternative would be SMALL.

7.1.8 Environmental Justice

Current operations at NBSR have no disproportionate impacts (adverse or otherwise) on the minority and low-income populations of Montgomery County and the surrounding counties, and no environmental pathways have been identified that would cause disproportionate impacts. Closure of the NBSR could result in a small decrease in employment opportunities with possible slight negative and disproportionate impacts on minority or low-income populations that would be temporarily offset by the labor needed to support decommissioning activities. However, the small number of employees working at the NBSR is negligible when compared to the number of employment opportunities in the surrounding area. Overall, the staff concludes the environmental justice impacts under the no-action alternative would be SMALL.

7.1.9 Aesthetics and Noise

Decommissioning would result in the eventual dismantlement of buildings and structures at the NIST site, resulting in a positive aesthetic impact. Noise that may be detectable offsite from the NIST campus would be generated during decommissioning operations; however, the impact is not likely to destabilize or alter any important attribute of the resource. Overall, the staff concludes the aesthetic and noise impacts associated with the no-action alternative would be SMALL.

7.1.10 Historic and Archaeological Resources

The amount of undisturbed land needed to support the decommissioning process would be relatively small. Decommissioning activities conducted on the NIST campus would not be expected to have a detectable effect on important cultural resources. The Maryland Inventory of Historic Properties does not have any records of known archaeological sites or other historic properties within or immediately adjacent at the NBSR or the entire NIST campus (MDP 2006). Nevertheless, in the event that any historic and archaeological resources on the NIST campus were discovered, these resources would not be expected to be adversely affected during decommissioning. It is likely that the NBSR wing of the 235 Building would be retained by the NIST following decommissioning. The staff concludes the impacts of the no-action alternative on historic and archaeological resources would be SMALL.

7.2 Construction of a Replacement Reactor and Associated Facilities

The alternative of constructing a replacement reactor and associated support facilities for the NBSR is discussed in this section. Under this alternative, it is assumed the OL for the NBSR would not be renewed and a new replacement reactor and associated support facilities would be constructed, perhaps, at another site. The support facilities are assumed to include a cooling tower, fuel storage area, a ventilation and exhaust stack, a facility comparable to the existing Cold Neutron Guide Hall, an office building, and a building for service equipment. The analysis is based on construction of a replacement reactor and associated facilities at some alternate location east of the Mississippi River; no specific site for new construction is assumed.

Some of the estimated impact information in Section 7.2 is adapted from a DOE EIS (U.S. DOE 2000). Section 4.6 of the DOE EIS evaluated the construction of a new research reactor at a generic DOE site for the production of plutonium-238, isotopes for medical and industrial uses, and materials testing for civilian nuclear energy research and development.

DOE currently supplies the uranium fuel used by the NBSR (NIST 2004). It is assumed DOE would also supply the fuel for a new replacement reactor.

The staff's characterizations of the impacts associated with construction and operation of a replacement reactor at an alternate location are shown in Table 7-2.

Table 7-2. Characterization of Impacts Associated with Construction and Operation of a Replacement Reactor and Associated Support Facilities

Impact Category	Construction	Operation
Land Use	SMALL	SMALL
Aquatic and Terrestrial Resources	SMALL	SMALL
Water Use and Quality	SMALL	SMALL
Air Quality	SMALL	SMALL
Waste	SMALL	SMALL
Human Health	SMALL	SMALL
Socioeconomics	SMALL	SMALL
Aesthetics and Noise	SMALL	SMALL
Historic and Archaeological Resources	SMALL	SMALL
Environmental Justice	SMALL	SMALL

7.2.1 Land Use

The construction of a new reactor and support facilities would disturb as much as approximately 4 ha (10 ac). It is assumed siting would be conducted so construction would be compatible with local zoning and the Coastal Zone Management Program if such a program is applicable in the hosting state. Clearing and grading operations could result in the loss of wetlands, although proper placement of the reactor and support facilities would eliminate or reduce the potential for such loss. Potential impacts on wetlands would be mitigated by the implementation of best management practices.

Overall, the staff concludes impacts on land use from constructing and operating a replacement reactor and associated support facilities would be SMALL.

7.2.2 Aquatic and Terrestrial Resources

During construction, impacts on aquatic resources could result from stormwater runoff. Runoff could alter flow rates, increase turbidity, and lead to sedimentation of streambeds. These impacts could, in turn, cause temporary and permanent changes in species composition and density and alter breeding habitats. Implementation of erosion and sediment control procedures would lessen construction impacts. Operational impacts on aquatic resources could occur as a result of water withdrawal and discharge. Water withdrawal could lead to the loss of aquatic organisms through impingement or entrainment. Discharge of cooling water could result in alterations in aquatic communities. Alterations could include changes in aquatic vegetation and the loss of fish and benthic macroinvertebrates. Additionally, radionuclides and chemicals in the discharge water have the potential to impact aquatic organisms. The extent of potential impacts on the aquatic environment would depend upon site- and facility-specific design information.

Construction of a replacement reactor and support facilities would likely result in the loss of woodland habitat at the alternate location. Land-clearing activities would affect animal populations. Less mobile animals within the project area, such as reptiles and small mammals, might not be expected to survive. Construction activities and noise would cause larger mammals and birds in the construction and adjacent areas to move to similar habitat nearby. If the area to which they moved was below its carrying capacity, these animals would be expected to survive. However, if the area were already supporting the maximum number of individuals, the additional animals would compete for limited resources that could lead to habitat degradation and eventual loss of the excess population. Nests and young animals living within the disturbed area might not survive. Activities associated with operations could affect wildlife living adjacent to the research reactor and support facilities. Emissions to the air and water, both nonradiological and radiological, could impact both plants and animals. Plants and animals could be exposed to pollutants via a number of pathways, including direct exposure, contact with contaminated soil, ingestion, and inhalation. Bioaccumulation could affect species that consume exposed plants or animals.

Construction and operation of a replacement reactor and support facilities could have the potential to impact threatened and endangered species. Consultations with the Fish and Wildlife Service, the Fisheries Service, and appropriate State agencies would be conducted at the site-specific level, as appropriate, to minimize adverse impacts.

Although the impacts on aquatic and terrestrial resources cannot be known with certainty given the assumption of siting at some alternate location, the staff estimates the aquatic and terrestrial resource impacts of constructing and operating a replacement reactor and associated support facilities at some alternate location east of the Mississippi River would be SMALL.

7.2.3 Water Use and Quality

During construction of a replacement reactor and support facilities, water is expected to be required for such uses as concrete mixing, dust control, washing activities, and potable and sanitary needs. The impact of these withdrawals on the resource would depend on the water source (surface water or groundwater) and its relative abundance. Impacts would be expected to be small because of the relatively small volumes of water required for construction compared to expected water availability. Nearby wastewater treatment facilities would be used to the extent possible and would be supplemented by portable or temporary facilities during construction as necessary. All wastewater would be disposed of in accordance with applicable regulatory requirements with discharges to surface waters in accordance with National Pollutant Discharge Elimination System (NPDES) effluent requirements. Ground disturbance and runoff from cleared areas could potentially impact surface water quality near construction areas. However, appropriate spill prevention practices and soil erosion and sediment control measures (e.g., use of silt fences and mulching and seeding disturbed areas) would be employed during construction to minimize water quality impacts.

During operation, water would be required to support such uses as process cooling and potable and sanitary needs. The single largest system use would be for cooling tower operation and associated evaporative losses. The impact of these withdrawals on the resource would depend on the water source (i.e., surface water or groundwater) and its relative abundance. For surface water, a dedicated surface water intake might have to be constructed if the site's existing distribution system is inadequate to meet the increased demands of the facilities. For ground-water, additional wells might have to be developed to supply the facilities directly or to provide increased production capacity for the site's existing supply system. It is expected that process effluent would mainly consist of cooling tower blowdown. There would be no radiological liquid effluent discharge to the environment under normal operations. Wastewater would be generated as a result of staff use of lavatories, showers, kitchens, and experimental facilities, and from miscellaneous potable and sanitary uses. Process and sanitary wastewater would be discharged to either existing site wastewater treatment facilities or to new facilities constructed specifically to serve the new reactor and support operations. All wastewater would be disposed of in accordance with applicable regulatory requirements with discharges to surface waters in accordance with NPDES effluent limitations.

Although the impacts on water use and quality cannot be known with certainty, assuming some alternate location east of the Mississippi River, the staff estimates the water use and quality impacts of constructing and operating a replacement reactor and associated support facilities would be SMALL.

7.2.4 Air Quality

Construction of a new reactor and support facilities would result in an increase in vehicle traffic with associated emissions. Some construction equipment would have emissions, and fugitive dust emissions from the construction process would also occur. During operation, emissions from the stack exhaust would be comparable to those for the NBSR and associated facilities. All construction and operation activities would be conducted in compliance with applicable regulatory requirements for air emissions.

Although the impacts on air quality cannot be known with certainty, assuming some alternate location east of the Mississippi River, the staff estimates the air quality impacts of constructing and operating a replacement reactor and associated support facilities would be SMALL provided the region is in attainment for National Ambient Air Quality Standards.

7.2.5 Waste

During construction, nonhazardous waste and debris would be generated. These materials would be disposed of offsite in disposal facilities with appropriate permits.

During operation, waste impacts would be comparable to those for the NBSR and associated facilities, as discussed in Chapter 5 of this EIS.

Overall, the staff estimates the waste impacts from constructing and operating a replacement reactor and associated facilities at a generic eastern site would be SMALL, but could be larger than continuing use of the current facility

7.2.6 Human Health

During construction of a replacement reactor and associated facilities, it is anticipated there would be no radiological health impacts beyond exposure to natural background levels in the construction area. Construction workers could experience industrial accidents that are possible at any construction activity.

During operation, human health impacts would be comparable to those for the NBSR, as discussed in Chapters 3, 4, and 5 of this EIS.

Overall, the staff estimates the human health impacts from constructing and operating a replacement reactor and associated facilities, assuming some alternate location east of the Mississippi River, would be SMALL, but could be larger than continuing operation of the current facility because of construction impacts.

7.2.7 Socioeconomics

It is estimated that on the order of 100 workers would be needed for a time period of 2 to 3 years to construct a replacement reactor and associated support facilities. The socioeconomic impacts of this workforce would be limited unless the site selected was in a remote, rural area.

During operation, socioeconomic impacts would be comparable to those of the NBSR, assuming location of the replacement reactor in an urban area. For location in a rural area, socioeconomic impacts could be somewhat greater although they would still be small given the limited workforce required to operate the reactor.

Although impacts cannot be known with certainty, assuming some alternate location east of the Mississippi River, the staff estimates the socioeconomic impacts of construction and operation of a replacement reactor and associated support facilities would be SMALL.

7.2.8 Environmental Justice

Construction and operation of a replacement reactor and associated support facilities would be unlikely to have disproportionately high and adverse health or environmental impacts on minority or low-income populations because radiological and nonradiological risks to persons residing in potentially affected areas would not be significant.

Overall, the staff estimates the environmental justice impacts from constructing and operating a replacement reactor and associated facilities at some alternate location east of the Mississippi River would be SMALL.

7.2.9 Aesthetics and Noise

Construction and operation of a replacement reactor and associated support facilities would have an aesthetic impact. The extent of the impact would depend on the location chosen and the surrounding land and land uses. The NBSR facility is housed in a building that is low to the ground; the staff assumes a replacement reactor would be similarly unobtrusive. The staff also assumed the cooling system for a replacement reactor would have a plume suppression cooling tower similar to that used for the NBSR (NIST 2004).

Construction of a replacement reactor and support facilities would result in some increase in noise levels from the use of earthmoving, materials-handling and impact equipment, employee vehicles, and truck traffic. Noise from construction activities, especially impulsive noise (e.g., jack hammers) would be temporary but could disturb wildlife in the immediate area of the construction site. The change in noise levels in areas outside the site would depend on the location selected and the exact nature of the construction location and activities required.

Operation of a replacement reactor and support facilities would result in some increase in noise levels from equipment (e.g., cooling systems, vents, motors, generators, compressors, pumps, and material-handling equipment), employee vehicles, and truck traffic. Noise from operation activities could disturb wildlife outside the facility fence line. The change in noise levels in areas outside the site would depend on the location selected, the size of the site, and the equipment used.

Overall, the staff estimates the aesthetic and noise impacts from constructing and operating a replacement reactor and associated facilities at some alternate location east of the Mississippi River would be SMALL.

7.2.10   Historic and Archaeological Resources

Because the exact nature of the site for a replacement reactor and associated support facilities is not known, potential effects of construction and operation on cultural resources cannot be determined. In general, if the alternate location had been previously developed, impacts on cultural resources might not occur. However, if an undisturbed location were chosen, cultural resources could be impacted. Historic and archaeological resources, including those that are or may be eligible for listing on the National Register of Historic Places, would be identified through site surveys and consultation with the State Historic Preservation Officer. Specific concerns about the presence, type, and location of Native American resources would be addressed through consultation with the potentially affected tribes in accordance with the National Historic Preservation Act, the Native American Graves Protection and Repatriation Act, and the American Indian Religious Freedom Act.

Although the impacts of construction and operation of a replacement reactor and associated support facilities on historic and archaeological resources cannot be known with certainty, assuming some alternate location east of the Mississippi River, the staff estimates the impacts would be SMALL.

7.3   Summary of Alternatives Considered

The adverse environmental impacts resulting from either of the alternatives considered by the staff if the NBSR ceases operation upon final determination of the license renewal application will not be smaller than those associated with continued operation, and they may be greater for some environmental issues in some locations.

7.4 References

10 CFR Part 20. Code of Federal Regulations, Title 10, *Energy*, Part 20, "Standards for Protection Against Radiation."

10 CFR Part 50. Code of Federal Regulations, Title 10, *Energy*, Part 50, "Domestic Licensing of Production and Utilization Facilities."

10 CFR Part 51. Code of Federal Regulations, Title 10, *Energy*, Part 51, "Environmental Protection Regulations for Domestic Licensing and Related Regulatory Functions."

American Indian Religious Freedom Act. 42 USC 1996, et seq.

Maryland Department of Planning (MDP). 2006. Letter, dated February 17, 2006, from Elizabeth J. Cole, Administrator, Project Review and Compliance to Katie A. Cort, PNNL. ML070380446.

National Environmental Policy Act of 1969 (NEPA). 42 USC 4321, et seq.

National Institute of Standards and Technology (NIST). 2004. *Environmental Report for License Renewal for the National Institute of Standards and Technology Reactor*. NBSR-16, NISTIR 7105, NIST, Gaithersburg, Maryland.

National Historic Preservation Act of 1966 (NHPA). 16 USC 470, et seq.

Native American Graves Protection and Repatriation Act. 25 USC 3001, et seq.

U.S. Department of Energy (U.S.DOE). 2000. *Final Programmatic Environmental Impact Statement for Accomplishing Expanded Civilian Nuclear Energy Research and Development and Isotope Production Missions in the United States, Including the Role of the Fast Flux Test Facility*. DOE/EIS-0310, Washington, D.C. Accessed on the Internet at http://www.eh.doe.gov/nepa/eis/eis0310/eis0310.html.

U.S. Nuclear Regulatory Commission (U.S.NRC). 1996. *Generic Environmental Impact Statement for License Renewal of Nuclear Plants*. NUREG-1437, Volumes 1 and 2, Washington, D.C.

U.S. Nuclear Regulatory Commission (U.S.NRC). 1999. *Generic Environmental Impact Statement for License Renewal of Nuclear Plants, Main Report*, "Section 6.3 – Transportation, Table 9.1, Summary of findings on NEPA issues for license renewal of nuclear power plants, Final Report." NUREG-1437, Volume 1, Addendum 1, Washington, D.C.

8.0 Summary and Conclusions

By letter dated April 9, 2004, the National Institute of Standards and Technology (NIST) submitted an application to the U.S. Nuclear Regulatory Commission (NRC) to renew the operating license (OL) for the National Bureau of Standards Reactor (NBSR) for an additional 20-year period (NIST 2004). If the OL is renewed, NIST and other decisionmakers will ultimately decide whether the reactor will continue to operate. If the OL is not renewed, then the reactor must be shut down upon NRC's determination of the application. The current OL for the NBSR was scheduled to expire on May 16, 2004. However, in accordance with 10 CFR 2.109(a), NIST's application for renewal was received at least 30 days prior to the expiration of the current license, and therefore, the existing OL will not be considered expired until the application has been finally determined.

Section 102 of the National Environmental Policy Act of 1969 (NEPA) (42 USC 4321, et seq.) directs that an environmental impact statement (EIS) is required for major Federal actions that significantly affect the quality of the human environment. The NRC has implemented Section 102 of NEPA in Title 10 of the Code of Federal Regulations (CFR) Part 51. Part 51 identifies licensing and regulatory actions that require an EIS. In 10 CFR 51.20(b)(2), the Commission requires preparation of an EIS for renewal of a testing facility (test reactor) OL.

Upon acceptance of the NIST application, the NRC began the environmental review process described in 10 CFR Part 51 by publishing a notice of intent to prepare an EIS and conduct scoping (70 FR 56935) on September 29, 2005. The staff visited the NIST site in September 2006. The staff reviewed the Environmental Report (ER) submitted by NIST (NIST 2004), consulted with other agencies, and conducted an independent analysis of the issues. No comments were received from the public during the scoping process in advance of the preparation of this EIS.

A Draft EIS was published for comment in June 2007. On July 18, 2007, the NRC published a Notice of Availability for the Draft EIS, thus initiating a comment period that ended on September 5, 2007 (72 FR 39467). Two comments – one from the U.S. Department of the Interior and one from the U.S. Environmental Protection Agency – were received, and these comments are addressed in Appendix B, Part II, of this Final EIS.

This EIS includes the NRC staff's analysis that considers and weighs the environmental effects of the proposed action, the environmental impacts of alternatives to the proposed action, and mitigation measures available for reducing or avoiding adverse effects. It also includes the staff's recommendation regarding the proposed action.

For this license renewal review, the NRC considered the purpose and need for the proposed action (i.e., renewal of the NBSR OL) is to provide an option allowing for neutron research capabilities beyond the term of the current reactor operating license to meet future national research and test facility needs, as such needs may be determined by NIST. There may be factors, in addition to NRC's license renewal determination, that will ultimately determine whether the NIST test reactor continues to operate beyond the determination of this license renewal action.

For the evaluation of the NBSR license renewal action, the staff has applied the NRC's three-level standard of significance – SMALL, MODERATE, or LARGE – developed using the Council on Environmental Quality guidelines. The following definitions of the three significance levels are set forth in the footnotes to Table B-1 of 10 CFR Part 51, Subpart A, Appendix B:

> SMALL – Environmental effects are not detectable or are so minor that they will neither destabilize nor noticeably alter any important attribute of the resource.

> MODERATE – Environmental effects are sufficient to alter noticeably, but not to destabilize, important attributes of the resource.

> LARGE – Environmental effects are clearly noticeable and are sufficient to destabilize important attributes of the resource.

The staff considered the environmental impacts associated with alternatives to license renewal and compared the environmental impacts of license renewal and the alternatives. The alternatives to license renewal that were considered include the no-action alternative (i.e., not renewing the OL for the NBSR) and replacement of the capabilities of the NBSR.

8.1 Environmental Impacts of the Proposed Action – License Renewal

The staff has established an independent process for identifying and evaluating the environmental impacts associated with license renewal. Neither the scoping process, NIST staff, nor the NRC staff has identified any issue applicable to the NBSR that would have a significant environmental impact. Measures were considered for mitigation of the environmental impacts of plant operation. The existing mitigation measures were found to be adequate, and no additional mitigation measures were deemed sufficiently beneficial to be warranted.

The following sections discuss unavoidable adverse impacts, irreversible or irretrievable commitments of resources, and the relationship between local short-term use of the environment and long-term productivity.

8.1.1 Unavoidable Adverse Impacts

An environmental review conducted at the license-renewal stage differs from the review conducted in support of a construction permit or initial OL because the plant is in existence at the license-renewal stage and has operated for a number of years. As a result, adverse impacts associated with the initial construction have been avoided, have been mitigated, or have already occurred. The environmental impacts to be evaluated for license renewal are those associated with continued operation during the renewal term; NIST did not consider that major refurbishment activities would be necessary for the continued operation of the NBSR.

The adverse impacts of continued operation identified are considered to be of SMALL significance, and none warrants implementation of additional mitigation measures. The staff concludes adverse impacts of likely alternatives if the NBSR ceases operation upon final determination of the licence renewal application will not be smaller than those associated with continued operation, and they may be greater for some environmental issues in some locations.

8.1.2 Irreversible or Irretrievable Resource Commitments

The commitment of resources related to construction and operation of the NBSR during the current license period was made when the plant was built. The resource commitments to be considered in this EIS are associated with continued operation of the plant for an additional 20 years. These resources include materials and equipment required for plant maintenance and operation, the nuclear fuel used by the reactor and, ultimately, disposition of the spent fuel assemblies.

The most significant resource commitments related to operation during the renewal term are the fuel and the permanent spent fuel disposition. NIST replaces 4 of the 30 fuel elements every refueling outage, which occurs at 5- to 6-week intervals.

If the NBSR ceases operation upon final determination of the current application, the likely alternative would require a commitment of resources for construction of a replacement reactor and test facility as well as for fuel to operate such a reactor.

8.1.3 Short-Term Use Versus Long-Term Productivity

An initial balance between short-term use and long-term productivity of the environment at the NIST site was set when construction of the NBSR was approved and construction began. That balance is now well established. Renewal of the OL for NBSR and continued operation of the reactor will not alter the existing balance, but may postpone the availability of that portion of the building complex housing the reactor for other uses. Denial of the application to renew the OL would lead to shutdown of the reactor and would alter the balance in a manner that would depend on subsequent uses of the building or the site.

8.2 Relative Significance of the Environmental Impacts of License Renewal and Alternatives

The proposed action is renewal of the OL for the NBSR. Chapter 2 describes the site, reactor, and interactions of the reactor with the environment. As noted in Chapter 3, no refurbishment and no refurbishment impacts are expected at the NBSR. Chapters 3 through 6 discuss environmental issues associated with renewal of the OL. Environmental issues associated with the no-action alternative and alternatives involving construction and operation of a replacement facility are discussed in Chapter 7.

The significance of the environmental impacts from the proposed action (i.e., approval of the application for renewal of the OL), the no-action alternative (i.e., denial of the application), and construction of new research capabilities at some alternate eastern location are listed in Table 8-1. Construction of facilities similar to the NBSR is assumed for the alternate location.

Table 8-1 shows the significance of the environmental effects of the proposed action are SMALL for all impact categories. The alternative actions, including the no-action alternative, may have environmental effects in at least some impact categories that, although considered SMALL, could be larger than the impacts of license renewal of the existing NBSR.

Table 8-1. Summary of Environmental Significance of License Renewal, the No-Action Alternative, and Construction and Operation of Alternative Research Facilities

Impact Category	Proposed Action License Renewal	No-Action Alternative Denial of Renewal	Replacement Facility
Land Use	SMALL	SMALL	SMALL
Ecology	SMALL	SMALL	SMALL
Water Use and Quality-Surface Water	SMALL	SMALL	SMALL
Water Use and Quality-Groundwater	SMALL	SMALL	SMALL
Air Quality	SMALL	SMALL	SMALL
Waste	SMALL	SMALL	SMALL
Human Health	SMALL	SMALL	SMALL
Socioeconomics	SMALL	SMALL	SMALL
Aesthetics	SMALL	SMALL	SMALL
Historic and Archaeological Resources	SMALL	SMALL	SMALL
Environmental Justice	SMALL	SMALL	SMALL

8.3 Staff Conclusions and Recommendations

Based on the ER submitted by NIST (NIST 2004); consultation with Federal, State, and local agencies; the staff's independent analysis; the opportunity to consider public comments during the scoping process, and comments received on the Draft EIS during the comment period, the preliminary recommendation of the staff is that the Commission determines the adverse environmental impacts of license renewal for NBSR are not so great that preserving the option of license renewal for Federal decision-makers would be unreasonable.

8.4 References

10 CFR Part 51. Code of Federal Regulations, Title 10, *Energy,* Part 51, "Environmental Protection Regulations for Domestic licensing and Related Regulatory Functions."

70 FR 56935. 2005. "National Institute of Standards and Technology, National Bureau of Standards Reactor; Notice of Intent to Prepare an Environmental Impact Statement and Conduct Scoping Process." *Federal Register.* Vol. 70, No. 188, pp. 56,935-56,936. September 29, 2005.

72 FR 39467. 2007. "National Institute of Standards and Technology; National Bureau of Standards Reactor; Notice of Availability of the Draft Environmental Impact Statement for License Renewal and Public Comment Period for the License Renewal of National Bureau of Standards Reactor." *Federal Register,* Vol. 72, No. 137, p. 39,467. July 18, 2007.

National Environmental Policy Act of 1969 (NEPA). 42 USC 4321, et seq.

National Institute of Standards and Technology (NIST). 2004. *Environmental Report for License Renewal for the National Institute of Standards and Technology Reactor.* NBSR-16, NISTIR 7105, NIST, Gaithersburg, Maryland.

Appendix A

Contributors to the Document

Appendix A

Contributors to the Document

The overall responsibility for the preparation of this environmental impact statement was assigned to the Office of Nuclear Reactor Regulation, U.S. Nuclear Regulatory Commission (NRC). The statement was prepared by members of the Office of Nuclear Reactor Regulation with assistance from other NRC organizations and the Pacific Northwest National Laboratory.

Name	Affiliation	Function or Expertise
Nuclear Regulatory Commission		
James Wilson	Nuclear Reactor Regulation	Project Management
Dennis Beissel	Nuclear Reactor Regulation	Project Management
Barry Zalcman	Nuclear Reactor Regulation	Project Management
Pacific Northwest National Laboratory[a]		
Beverly Miller		Task Leader
Eva Eckert Hickey		Deputy Task Leader
Jeremy Rishel		Air Quality
Katherine Cort		Socioeconomics and Cultural Resources
Amanda Stegen		Aquatic and Terrestrial Ecology
Kathleen Rhoads		Radiation Protection
Paul Hendrickson		Land Use, Related Federal Programs, Alternatives
Lance Vail		Water Use, Hydrology
James Weber		Technical Editor
Cary Counts		Technical Editor
Dave Payson		Technical Editor
Lila Andor		Document Production
Susan Tackett		Document Production
Mike Parker		Document Production

(a) Pacific Northwest National Laboratory is operated for the U.S. Department of Energy by Battelle Memorial Institute.

Appendix B

Comments Received on the Environmental Review

Appendix B

Comments Received on the Environmental Review

Part I – Comments Received During Scoping

On September 29, 2005, the U.S. Nuclear Regulatory Commission (NRC) published a Notice of Intent in the *Federal Register* (70 FR 56935) to notify the public of the staff's intent to prepare a plant-specific environmental impact statement (EIS) to consider the renewal application for the National Institute of Standards and Technology National Bureau of Standards Reactor operating license and to conduct scoping. This EIS has been prepared in accordance with the National Environmental Policy Act of 1969, and Title 10 of the Code of Federal Regulations (CFR) Part 51. As outlined by 10 CFR Part 51, the NRC initiated the scoping process with the issuance of the *Federal Register* Notice. The NRC invited the applicant; Federal, State, Native American Tribal, and local government agencies; local organizations; and individuals to participate in the scoping process by submitting written suggestions and comments no later than November 28, 2005. No comments were received during the scoping period.

Part II – Comments Received on the Draft EIS

Pursuant to 10 CFR Part 51, the staff transmitted the *Draft Environmental Impact Statement for License Renewal of the National Bureau of Standards Reactor, Draft Report for Comment* to Federal, State, Native American Tribal, and local government agencies as well as interested members of the public. As part of the process to solicit comments on the Draft EIS, the staff:

- placed a copy of the Draft EIS in the NRC's electronic Public Document Room and its license renewal website

- sent copies of the Draft EIS to the applicant and certain Federal, State, and local agencies

- published a Notice of Availability of the Draft EIS in the *Federal Register* on July 18, 2007 (72 FR 39467)

- issued public service announcements and instructions on how to comment on the Draft EIS

- established a website to receive comments on the Draft EIS through the Internet.

Appendix B

During the comment period, the staff received a total of two comment letters.

After the comment period expired, staff reviewed the two comment letters that are part of the docket file for the application, both of which are available in the NRC's electronic Public Document Room. Appendix B, Part II, Section B.2, contains a summary of the comments and the staff's response. Appendix B, Part II, Section B.3, contains copies of the comment letters.

Each comment identified by the staff was assigned a specific alpha-numeric identifier. A cross-reference of the alpha-numeric identifiers, the author of the comment, the page where the comment can be found, and the section(s) of this report in which the comment is addressed is provided in Table B-1. The written comment letters are identified by the letters A and B. The accession number is provided for the written comments to facilitate access to the document through the Public Electronic Reading Room (ADAMS) http://www.nrc.gov/reading-rm/adams/login.html.

Table B-1. Comments Received on the Draft EIS

Comment No.	Author	Source	Page of Comment	Section(s) Where Addressed
A-01	Michael Chezik U.S. Department of the Interior	Letter, August 29, 2007 ML072530657	B-2	B.2.1
B-02	William Arguto U.S. Environmental Protection Agency	Letter, September 5, 2007 ML072560366	B-2	B.2.1

B.1 Comments and Responses on the Draft EIS

Two comment letters were received during the review period for the NBSR Draft EIS. Both of these letters are general in nature, and neither identify issues that require further consideration by the NRC staff. The comments are listed in Section B.2.1 along with the NRC response. Copies of the letters are provided in Section B.2.2.

B.2.1 General Comments Concerning License Renewal Process

Comment: The U.S. Department of the Interior has no comment on the Draft Environmental Impact Statement for License Renewal of the National Bureau of Standards Reactor (Draft NUREG-1873), published June 2007. (A-01)

Comment: Under EPA's system for rating Environmental Impact Statements, we are rating the environmental impacts associated with the operating license renewal as a Lack of Objections (LO-1). A Lack of Objection rating means the review has not identified any potential environmental impacts requiring substantive changes to the preferred alternative. The numeric rating assesses the adequacy of the Environmental Impact Statement. The 1 rating indicates DEIS adequacy sets forth the environmental impact(s) of the preferred alternative and those of the alternatives reasonably available to the project or action. No further analysis or data collection is necessary. A copy of our rating system is attached, and can also be found at : http://www.epa.gov/Compliance/nepa/comments/ratings/html. (B-01)

Response: *These comments are acknowledged and did not provide significant new information relevant to this EIS. Therefore, they will not be evaluated further. There were no changes made in this EIS as a result of these comments.*

B.2.1 Letters Received on the Draft EIS

United States Department of the Interior

OFFICE OF THE SECRETARY
Office of Environmental Policy and Compliance
Custom House, Room 244
200 Chestnut Street
Philadelphia, Pennsylvania 19106-2904

IN REPLY REFER TO:

TAKE PRIDE
IN AMERICA

August 29, 2007

RECEIVED

'07 SEP -4 PM 2: 45

RULES ... DIRECTIVES

ER 07/606

Chief, Rulemaking, Directives and Editing Branch
Mail Stop: T6-D59
U.S. Nuclear Regulatory Commission
Washington, DC 20555-0001

Dear Sir/Madam:

The U.S. Department of the Interior has no comment on the Draft Environmental Impact   A-01
Statement (DEIS) for License Renewal of the National Bureau of Standards Reactor
(Draft NUREG-1873), published June 2007.

Thank you for the opportunity to review and comment on the DEIS.

7/18/07
72FR 39467
(1)

Sincerely,

Michael T. Chezik
Regional Environmental Officer

E-REDS = ADM-03
all = D. Beissel (DRB1)

SONSI Review Complete
Template = ADM-013

UNITED STATES ENVIRONMENTAL PROTECTION AGENCY
REGION III
1650 Arch Street
Philadelphia, Pennsylvania 19103-2029

September 5, 2007

Mr. Dennis Beissel, Project Manager   *7/18/07*
Mail Stop: T6-D59
U.S. Nuclear Regulatory Commission   *72 FR 39467*
Washington, DC 20555-0001   ②

RECEIVED

RE: Comments to the Draft Environmental Impact Statement for License Renewal
of the National Bureau of Standards Reactor – NUREG-1873, CEQ # 20060290.

Dear Mr. Beissel:

    In accordance with the National Environmental Policy Act (NEPA), Section 309
of the Clean Air Act, and the Council on Environmental Quality (CEQ) regulations
implementing NEPA (40 CFR 1500-1508), the U. S. Environmental Protection Agency
(EPA) has reviewed the Draft Environmental Impact Statement (DEIS) for the above
referenced project. The DEIS is for renewal of the operating license for the National
Bureau of Standards. The renewal would be for additional 20-year period.

    Under EPA's system for rating Environmental Impact Statements, we are rating
the environmental impacts associated with the operating license renewal as a Lack of
Objections (LO -1). A Lack of Objection rating means the review has not identified any
potential environmental impacts requiring substantive changes to the preferred
alternative. The numeric rating assesses the adequacy of the Environmental Impact
Statement. The 1 rating indicates the DEIS adequately sets forth the environmental
impact(s) of the preferred alternative and those of the alternatives reasonably available to
the project or action. No further analysis or data collection is necessary. A copy of our
rating system is attached, and can also be found at:
http://www.epa.gov/Compliance/nepa/comments/ratings.html.

B-01

    If you any questions regarding this issue please feel free to contact Kevin Magerr
at (215) 814-5724.

Sincerely,

William Arguto,
NEPA Team Leader

Attachment:

EREDS= ADM 03
Cel = D. Beissel (drb1)

SONSI Review Complete
Template = ADM-013

Appendix B

U.S. Environmental Protection Agency

National Environmental Policy Act (NEPA)

Recent Additions | Contact Us | Print Version   Search: ▓▓ Advanced Search

EPA Home > Compliance and Enforcement > National Environmental Policy Act (NEPA) > EPA Comments on Environmental
Impact Statements (EISs) > EIS Rating System Criteria

Compliance and
Enforcement Home

National
Environmental
Policy Act Home

Basic Information

Where You Live

Newsroom

Environmental Impact
Statement Database

Submitting
Environmental
Impact Statements

Obtaining
Environmental
Impact Statements

EPA Compliance with
NEPA

Environmental Impact Statement (EIS) Rating System Criteria

EPA has developed a set of criteria for rating draft EISs. The rating system provides a basis upon
which EPA makes recommendations to the lead agency for improving the draft EIS.

- Rating the Environmental Impact of the Action

- Rating the Adequacy of the Draft Environmental Impact Statement (EIS)

RATING THE ENVIRONMENTAL IMPACT OF THE ACTION

- **LO (Lack of Objections)** The review has not identified any potential environmental impacts
  requiring substantive changes to the preferred alternative. The review may have disclosed
  opportunities for application of mitigation measures that could be accomplished with no more
  than minor changes to the proposed action.

- **EC (Environmental Concerns)** The review has identified environmental impacts that should
  be avoided in order to fully protect the environment. Corrective measures may require changes
  to the preferred alternative or application of mitigation measures that can reduce the
  environmental impact.

- **EO (Environmental Objections)** The review has identified significant environmental impacts
  that should be avoided in order to adequately protect the environment. Corrective measures
  may require substantial changes to the preferred alternative or consideration of some other
  project alternative (including the no action alternative or a new alternative). The basis for
  environmental Objections can include situations:

  1. *Where an action might violate or be inconsistent with achievement or maintenance of a
     national environmental standard;*

  2. *Where the Federal agency violates its own substantive environmental requirements that
     relate to EPA's areas of jurisdiction or expertise;*

  3. *Where there is a violation of an EPA policy declaration;*

  4. *Where there are no applicable standards or where applicable standards will not be
     violated but there is potential for significant environmental degradation that could be
     corrected by project modification or other feasible alternatives; or*

  5. *Where proceeding with the proposed action would set a precedent for future actions that
     collectively could result in significant environmental impacts.*

- **EU (Environmentally Unsatisfactory)** The review has identified adverse environmental
  impacts that are of sufficient magnitude that EPA believes the proposed action must not
  proceed as proposed. The basis for an environmentally unsatisfactory determination consists of
  identification of environmentally objectionable impacts as defined above and one or more of the
  following conditions:

  1. *The potential violation of or inconsistency with a national environmental standard is
     substantive and/or will occur on a long-term basis;*

ttp://www.epa.gov/compliance/nepa/comments/ratings.html   9/5/2

2. There are no applicable standards but the severity, duration, or geographical scope of the impacts associated with the proposed action warrant special attention; or

3. The potential environmental impacts resulting from the proposed action are of national importance because of the threat to national environmental resources or to environmental policies.

Top of Page

RATING THE ADEQUACY OF THE DRAFT ENVIRONMENTAL IMPACT STATEMENT (EIS)

- **1 (Adequate)** The draft EIS adequately sets forth the environmental impact(s) of the preferred alternative and those of the alternatives reasonably available to the project or action. No further analysis or data collection is necessary, but the reviewer may suggest the addition of clarifying language or information.

- **2 (Insufficient Information)** The draft EIS does not contain sufficient information to fully assess environmental impacts that should be avoided in order to fully protect the environment, or the reviewer has identified new reasonably available alternatives that are within the spectrum of alternatives analyzed in the draft EIS, which could reduce the environmental impacts of the proposal. The identified additional information, data, analyses, or discussion should be included in the final EIS.

- **3 (Inadequate)** The draft EIS does not adequately assess the potentially significant environmental impacts of the proposal, or the reviewer has identified new, reasonably available, alternatives, that are outside of the spectrum of alternatives analyzed in the draft EIS, which should be analyzed in order to reduce the potentially significant environmental impacts. The identified additional information, data, analyses, or discussions are of such a magnitude that they should have full public review at a draft stage. This rating indicates EPA's belief that the draft EIS does not meet the purposes of NEPA and/or the Section 309 review, and thus should be formally revised and made available for public comment in a supplemental or revised draft EIS.

Appendix C

Chronology of NRC Staff Environmental Review Correspondence
Related to National Institute of Standards and Technology's
Application for License Renewal for the
National Bureau of Standards Reactor

Appendix C

Chronology of NRC Staff Environmental Review Correspondence
Related to National Institute of Standards and Technology's
Application for License Renewal for the
National Bureau of Standards Reactor

This appendix contains a chronological listing of correspondence between the U.S. Nuclear Regulatory Commission (NRC) and the National Institute of Standards and Technology (NIST), and other correspondence related to the NRC staff's environmental review, under Title 10 of the Code of Federal Regulations (CFR) Part 51, of NIST's application for a renewed operating license for the National Bureau of Standards Reactor (NBSR) on the NIST campus near Gaithersburg, Maryland. All documents, with the exception of those containing proprietary information, have been placed in the Commission's Public Document Room, at One White Flint North, 11555 Rockville Pike (first floor), Rockville, Maryland, and are available electronically from the Public Electronic Reading Room found on the Internet at the following web address: http://www.nrc.gov/reading-rm.html. From this site, the public can gain access to the NRC's Agencywide Documents Access and Management System (ADAMS), which provides text and image files of NRC's public documents in the Publicly Available Records component of ADAMS. The ADAMS accession numbers or *Federal Register* citation for each document are included below.

April 9, 2004	Letter from the National Institute of Standards and Technology (NIST) to NRC, regarding license renewal application for the National Bureau of Standards Reactor (NBSR) (Accession No. ML041120167), and Environmental Report (Accession No. ML041120176).
September 2, 2004	Letter from NRC to S. Weiss, NIST, regarding determination of acceptability and sufficiency for docketing, proposed review schedule, and opportunity for a hearing regarding the application from NIST for the NBSR. (Accession No. ML041390017)

Appendix C

September 21, 2004   NRC *Federal Register* Notice: National Institute of Standards and Technology, National Bureau of Standards (NIST); Notice of acceptance for docketing of the application and Notice of opportunity for hearing regarding renewal of the National Bureau of Standards reactor (NBSR) facility operating license No. TR-5 for an additional 20-year period. (69 FR 56462)

September 23, 2005   Letter from NRC to S. Weiss, NIST, transmitting notice of intent to prepare an environmental impact statement and conduct scoping. (Accession No. ML052660195)

September 29, 2005   NRC *Federal Register* Notice: National Institute of Standards and Technology, National Bureau of Standards Reactor; Notice of Intent to Prepare an Environmental Impact Statement and Conduct Scoping Process. (70 FR 56935)

February 17, 2006   Letter from NRC to W. Richards, NIST, regarding issuance of environmental scoping summary report associated with the staff's review of the application by the National Institute of Standards and Technology for renewal of the operating license for the National Bureau of Standards Reactor. (Accession No. ML032731680)

February 13, 2007   Letter from NRC to W. Richards, NIST, regarding summary of site audit to support the license renewal review for the NBSR at NIST. (Accession No. ML070370061)

April 3, 2007   Letter from NRC to Fish and Wildlife Service regarding consultation for protected species. (Accession No. ML07050245)

July 18, 2007   NRC *Federal Register* Notice: National Institute of Standards and Technology; National Bureau of Standards Reactor; Notice of Availability of the Draft Environmental Impact Statement for License Renewal and Public Comment period for the License Renewal of National Bureau of Standards Reactor. (72 FR 39467)

Appendix D

Organizations Contacted

Appendix D

Organizations Contacted

During the course of the NRC staff's independent analysis of environmental impacts from operations during the renewal term, the following Federal, State, regional, and local agencies were contacted:

Maryland Department of Natural Resources, Annapolis, Maryland

Maryland Historical Trust, Crownsville, Maryland

U.S. Fish and Wildlife Service, Annapolis, Maryland

NRC FORM 335 (9-2004) NRCMD 3.7	U.S. NUCLEAR REGULATORY COMMISSION	1. REPORT NUMBER (Assigned by NRC, Add Vol., Supp., Rev., and Addendum Numbers, if any.)
BIBLIOGRAPHIC DATA SHEET *(See instructions on the reverse)*		NUREG-1873

2. TITLE AND SUBTITLE	3. DATE REPORT PUBLISHED	
Environmental Impact Statement for License Renewal of the National Bureau of Standards Reactor	MONTH	YEAR
	December	2007
	4. FIN OR GRANT NUMBER	

5. AUTHOR(S)	6. TYPE OF REPORT
James Wilson Dennis Beissel Barry Zalcman	Final
	7. PERIOD COVERED *(Inclusive Dates)*
	1985-2007

8. PERFORMING ORGANIZATION - NAME AND ADDRESS *(If NRC, provide Division, Office or Region, U.S. Nuclear Regulatory Commission, and mailing address; if contractor, provide name and mailing address.)*

Pacific Northwest National Laboratory

Post Office Box 999

Richland, WA 99352

9. SPONSORING ORGANIZATION - NAME AND ADDRESS *(If NRC, type "Same as above"; if contractor, provide NRC Division, Office or Region, U.S. Nuclear Regulatory Commission, and mailing address.)*

Division of License Renewal

Office of Nuclear Reactor Regulation

U.S. Nuclear Regulatory Commission

Washington, D.C. 20555-0001

10. SUPPLEMENTARY NOTES

Docket No. 50-184

11. ABSTRACT *(200 words or less)*

This environmental impact statement (EIS) has been prepared in response to an application submitted to the U.S. Nuclear Regulatory Commission (NRC) by the National Institute of Standards and Technology (NIST) to renew the operating license for the National Bureau of Standards Reactor (NBSR) for a period of an additional 20 years. This is the second license renewal application for the NBSR. The first license renewal was granted May 16, 1984, and included a power uprate from 10 megawatts (MW) to 20 MW of thermal power. This EIS includes the NRC staff's analysis that considers and weighs the environmental impacts of the proposed action, as well as mitigation measures available for reducing or avoiding adverse impacts. It also includes the staff's recommendation regarding the proposed action.

No public comments were received during the scoping or review process. The staff determined from its review of the application that no issues having a significant environmental impact exist, and additional mitigation measures are not likely to be sufficiently beneficial as to be warranted.

The NRC staff's recommendation is that the Commission determine the adverse environmental impacts of license renewal for the NBSR are not so great that license renewal would be unreasonable. This recommendation is based on (1) the Environmental Report submitted by NIST; (2) consultation with Federal, State, and local agencies; and (3) the staff's own independent review.

12. KEY WORDS/DESCRIPTORS *(List words or phrases that will assist researchers in locating the report.)*	13. AVAILABILITY STATEMENT
EIS, NIST, NBSR, National Bureau of Standards, National Institute of Standards and Technology, test reactor, located in Maryland, license renewal	unlimited
	14. SECURITY CLASSIFICATION
	(This Page) unclassified
	(This Report) unclassified
	15. NUMBER OF PAGES
	16. PRICE